CONTROL THEORY FUNDAMENTALS

Edited by

Laurence Buen Suarez

Delve Publishing

Control Theory Fundamentals

Edited by Laurence Buen Suarez

Edition 2018

Delve Publishing

Published by
Delve Publishing LLC
708 3rd Avenue, 6th Floor
New York
NY 10801
United States of America
www.delvepublishing.com

ISBN: 978-1-68095-681-8 (Hardback)

Printed in United States of America

Library of Congress Control Number: 2016920455

About the Editor

Laurence Buen Suarez

Laurence is a licensed Instrumentation and Control Engineer graduated from Rizal Technological University. He ranked 6th in the certification exam (year 2012) and is a member of Philippine Instrumentation and Control Society (PICS). Currently, he is working as an Technical Engineer / Drafter and is providing documentation drawings for clients with projects related to different process industries.

Table of Contents

2 DYNAMICAL SYSTEM

3 CONTROLLER

4 PROCESS CONTROL

5 APPLICATIONS OF CONTROL THEORY

6 CLASSICAL CONTROL THEORY

7 LINEAR CONTROL SYSTEM

8 NONLINEAR CONTROL

9 DISTRIBUTED CONTROL SYSTEM

10 STOCHASTIC CONTROL

List of Abbreviations

APC	Advanced Process Control
ARC	Advanced Regulatory Control
ABS	Anti-lock Braking System
EBD	Electronic Brakeforce Distribution
COTS	Commercial Off-the-shelf
CT	Continuous Time
CBC	Cornering Brake Control
I/O	Input/Output
ISS	Input-to-state Stability
KAM	Kolmogorov–Arnold–Moser
DT	Discrete Time
DCS	Distributed Control System
LQG	Linear-quadratic-Gaussian control
LTI	Linear Time Invariant
MV	Manipulated Variable
MPO	Manufacturing Planning and Optimization
MCMC	Markov Chain Monte Carlo
MIMO	Multiple-input and Multiple-output
MPC	Multivariable Model Predictive Control
QFT	Quantitative Feedback Theory
PLC	Programmable Logic Controller
SISO	Single-input and Single-output
SMC	Sliding Mode Control

SPC	Statistical Process Control
TPMS	Tire Pressure Monitoring System
TCS	Traction Control System

Preface

Control theory is an interdisciplinary branch of engineering and mathematics that deals with the behavior of dynamical systems with inputs, and how their behavior is modified by feedback. The usual objective of control theory is to control a system, often called the plant, so its output follows a desired control signal, called the reference, which may be a fixed or changing value. To do this a controller is designed, which monitors the output and compares it with the reference. The difference between actual and desired output, called the error signal, is applied as feedback to the input of the system, to bring the actual output closer to the reference. Some topics studied in control theory are stability, controllability and observability. Although control theory has deep connections with classical areas of mathematics, such as the calculus of variations and the theory of differential equations, it did not become a field in its own right until the late 1950s and early 1960s. At that time, problems arising in engineering and economics were recognized as variants of problems in differential equations and in the calculus of variations, though they were not covered by existing theories. At first, special modifications of classical techniques and theories were devised to solve individual problems. It was then recognized that these seemingly diverse problems all had the same mathematical structure, and control theory emerged. Extensive use is usually made of a diagrammatic style known as the block diagram. The transfer function, also known as the system function or network function, is a mathematical representation of the relation between the input and output based on the differential equations describing the system. In recent years, control systems plays main role in the development and advancement of modern technology and civilization. Practically every aspects of our day-to-day life is affected less or more by some control system. A bathroom toilet tank, a refrigerator, an air conditioner, a geezer, an automatic iron, an automobile all are control system. These systems are also used in industrial process for more output. We find control system in quality control of products, weapons system, transportation systems, power system, space technology, robotics and many more. The principles of control theory is applicable to engineering and non-engineering field both.

Content Coverage

Chapters One and *Two* focus on introduction to control theory and study of dynamical systems, respectively. Dynamical system is a system in which a function describes the time dependence of a point in a geometrical space.

Chapters Three and *Four* present coverage of controller types and process control systems, separately. Typical applications of controllers are to hold settings for temperature, pressure, flow or speed, although process control enables automation, by which a small staff of operating personnel can operate a complex process from a central control room.

Chapters Five and *Six* focus on applications of control theory and classical control theory, respectively. Classical control theory deals with the behavior of dynamical systems with inputs, and how their behavior is modified by feedback, using the Laplace transform as a basic tool to model such systems.

Chapter Seven and *Eight* presents about linear control system and non-linear control system, distinctly. Linear control systems use linear negative feedback to produce a control signal mathematically based on other variables, with a view to maintain the controlled process within an acceptable operating range, but nonlinear control deals with systems that are nonlinear, time-variant, or both.

Chapters Nine and *Ten* present coverage of distributed control system (DCS) and stochastic optimal control, correspondingly. The DCS concept increases reliability and reduces installation costs by localising control functions near the process plant, but enables monitoring and supervisory control of the process remotely, however stochastic optimal control deals with the existence of uncertainty either in observations or in the noise that drives the evolution of the system.

CHAPTER 1

Control Theory

Chapter Contents

- **Introduction**
- **Overview**
- **Open-Loop And Closed-Loop (Feedback) Control**
- **Classical Control Theory**
- **Closed-Loop Transfer Function**
- **Pid Feedback Control**
- **Linear And Nonlinear Control Theory**
- **Analysis Techniques - Frequency Domain And Time Domain**
- **System Interfacing - Siso & Mimo**
- **System Classifications**
- **History**

Aim of this chapter is to introduce the notion of system. In the (abstract) definition of system, we follow a simple and natural approach, the so-called input-output approach, which is motivated by the study of simple systems. Among the various mathematical representations we give particular emphasis to the one based on the introduction of an auxiliary variable, the state variable, which is denoted as state space representation. This representation plays a fundamental role in systems and control theory, hence we discuss in detail its properties and the conditions under which it can be introduced.

INTRODUCTION

Control theory is an interdisciplinary branch of engineering and mathematics that deals with the behavior of dynamical systems with inputs, and how their behavior is modified by feedback. The usual objective of control theory is to control a system, often called the *plant*, so its output follows a desired control signal, called the *reference*, which may be a fixed or changing value. To do this a *controller* is designed, which monitors the output and compares it with the reference. The difference between actual and desired output, called the *error* signal, is applied as feedback to the input of the system, to bring the actual output closer to the reference. Some topics studied in control theory are stability (whether the output will converge to the reference value or oscillate about it), controllability and observability.

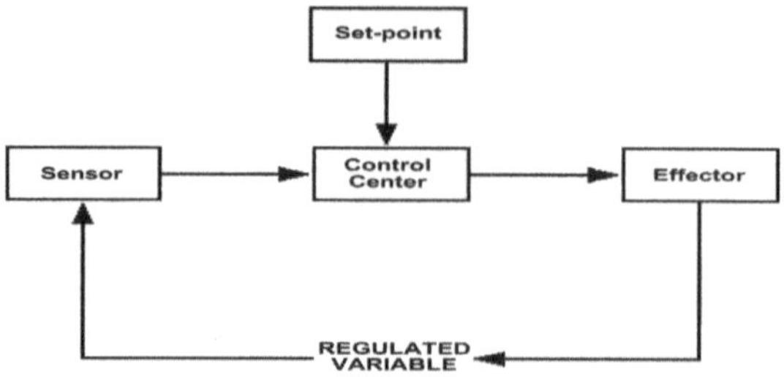

A block diagram of a negative feedback control system. Illustrates the concept of using a feedback loop to control the behavior of a system by comparing its output to a desired value, and applying the difference as an error signal to dynamically change the output so it is closer to the desired output

Extensive use is usually made of a diagrammatic style known as the block diagram. The transfer function, also known as the system function or network function, is a mathematical representation of the relation between the input and output based on the differential equations describing the system.

Although a major application of control theory is in control systems engineering, which deals with the design of process control systems for industry, other applications range far beyond this. As the general theory of feedback systems, control theory is useful wherever feedback occurs. A few examples are in physiology, electronics, climate modeling, machine design, ecosystems, navigation, neural networks, predator–prey interaction, gene expression, and production theory.

OVERVIEW

Smooth nonlinear trajectory planning with linear quadratic Gaussian feedback (LQR) control on a dual pendula system.

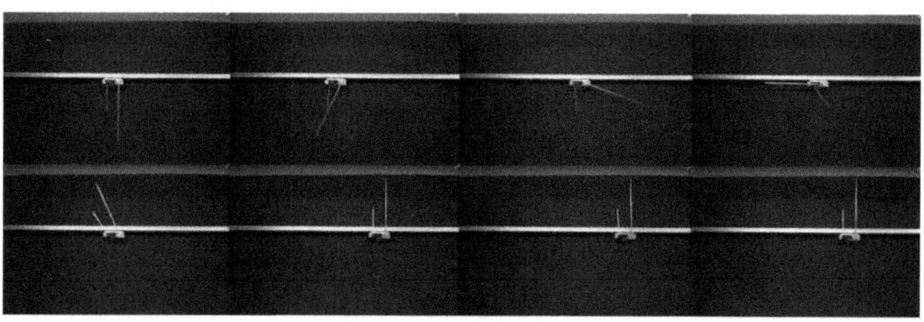

Control theory is

- a theory that deals with influencing the behavior of dynamical systems,
- an interdisciplinary subfield of science, which originated in engineering

and mathematics, and evolved into use by the social sciences, such as economics, psychology, sociology, criminology and in the financial system.

Control systems may be thought of as having four functions: measure, compare, compute and correct These four functions are completed by five elements: detector, transducer, transmitter, controller and final control element. The measuring function is completed by the detector, transducer and transmitter. In practical applications these three elements are typically contained in one unit. A standard example of a measuring unit is a resistance thermometer. The compare and compute functions are completed within the controller, which may be implemented electronically by proportional control, a PI controller, PID controller, bistable, hysteretic control or programmable logic controller. Older controller units have been mechanical, as in a centrifugal governor or a carburetor. The correct function is completed with a final control element. The final control element changes an input or output in the control system that affects the manipulated or controlled variable.

OPEN-LOOP AND CLOSED-LOOP (FEEDBACK) CONTROL

Fundamentally, there are two types of control loop; open loop control, and closed loop (feedback) control.

In open loop control, the control action from the controller is independent of the "process output" (or "controlled process variable"). A good example of this is a central heating boiler controlled only by a timer, so that heat is applied for a constant time, regardless of the temperature of the building. (The control action is the switching on/off of the boiler. The process output is the building temperature).

In closed loop control, the control action from the controller is dependent on the process output. In the case of the boiler analogy this would include a thermostat to monitor the building temperature, and thereby feed back a signal to ensure the controller maintains the building at the temperature set on the thermostat. A closed loop controller therefore has a feedback loop which ensures the controller exerts a control action to give a process output the same as the "Reference input" or "set point". For this reason, closed loop controllers are also called feedback controllers.

The definition of a closed loop control system according to the British Standard Institution is 'a control system possessing monitoring feedback, the deviation signal formed as a result of this feedback being used to control the action of a final control element in such a way as to tend to reduce the deviation to zero.' "

Likewise; "A *Feedback Control System* is a system which tends to maintain a prescribed relationship of one system variable to another by comparing functions of these variables and using the difference as a means of control.'"

The advanced type of automation that revolutionized manufacturing, aircraft, communications and other industries, is feedback control, which is usually *continuous* and involves taking measurements using a sensor and making calculated adjustments to keep the measured variable within a set range by means of a "final control element", such as a control valve.

The theoretical basis of closed loop automation is control theory.

Other Examples

An example of a control system is a car's cruise control, which is a device designed to maintain vehicle speed at a constant *desired* or *reference* speed provided by the driver. The *controller* is the cruise control, the *plant* is the car, and the *system* is the car and the cruise control. The system output is the car's speed, and the control itself is the engine's throttle position which determines how much power the engine delivers.

A primitive way to implement cruise control is simply to lock the throttle position when the driver engages cruise control. However, if the cruise control is engaged on a stretch of flat road, then the car will travel slower going uphill and faster when going downhill. This type of controller is called an *open-loop controller* because there is no feedback; no measurement of the system output (the car's speed) is used to alter the control (the throttle position.) As a result, the controller cannot compensate for changes acting on the car, like a change in the slope of the road.

In a *closed-loop control system*, data from a sensor monitoring the car's speed (the system output) enters a controller which continuously compares the quantity representing the speed with the reference quantity representing the desired speed. The difference, called the error, determines the throttle position (the control). The result is to match the car's speed to the reference speed (maintain the desired system output). Now, when the car goes uphill, the difference between the input (the sensed speed) and the reference continuously determines the throttle position. As the sensed speed drops below the reference, the difference increases, the throttle opens, and engine power increases, speeding up the vehicle. In this way, the controller dynamically counteracts changes to the car's speed. The central idea of these control systems is the *feedback loop*, the controller affects the system output, which in turn is measured and fed back to the controller.

CLASSICAL CONTROL THEORY

To overcome the limitations of the open-loop controller, control theory introduces feedback. A closed-loop controller uses feedback to control states or outputs of a dynamical system. Its name comes from the information path in the system: process inputs (e.g., voltage applied to an electric motor) have an effect on the process outputs (e.g., speed or torque of the motor), which is measured with sensors and processed by the controller; the result (the control signal) is "fed back" as input to the process, closing the loop.

Closed-loop controllers have the following advantages over open-loop controllers:

- disturbance rejection (such as hills in the cruise control example above)
- guaranteed performance even with model uncertainties, when the model structure does not match perfectly the real process and the model parameters are not exact
- unstable processes can be stabilized
- reduced sensitivity to parameter variations
- improved reference tracking performance

In some systems, closed-loop and open-loop control are used simultaneously. In such systems, the open-loop control is termed feedforward and serves to further improve reference tracking performance.

A common closed-loop controller architecture is the PID controller.

CLOSED-LOOP TRANSFER FUNCTION

The output of the system *y(t)* is fed back through a sensor measurement *F* to a comparison with the reference value *r(t)*. The controller *C* then takes the error *e* (difference) between the reference and the output to change the inputs *u* to the system under control *P*. This is shown in the figure. This kind of controller is a closed-loop controller or feedback controller.

This is called a single-input-single-output (*SISO*) control system; *MIMO* (i.e., Multi-Input-Multi-Output) systems, with more than one input/output, are common. In such cases variables are represented through vectors instead of simple scalar values. For some distributed parameter systems the vectors may be infinite-dimensional (typically functions).

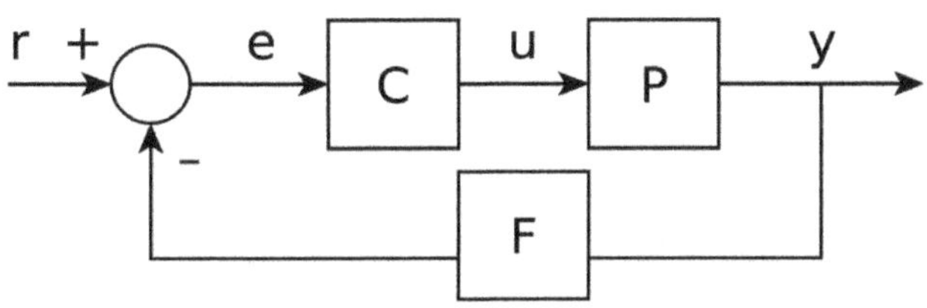

If we assume the controller *C*, the plant *P*, and the sensor *F* are linear and time-invariant (i.e., elements of their transfer function *C(s)*, *P(s)*, and *F(s)* do not depend on time), the systems above can be analysed using the Laplace transform on the variables. This gives the following relations:

$$Y(s) = P(s)U(s)$$
$$U(s) = C(s)E(s)$$
$$E(s) = R(s) - F(s)Y(s).$$

Solving for $Y(s)$ in terms of $R(s)$ gives

$$Y(s) = \left(\frac{P(s)C(s)}{1 + F(s)P(s)C(s)}\right) R(s) = H(s)R(s).$$

The expression $H(s) = \frac{P(s)C(s)}{1 + F(s)P(s)C(s)}$ is referred to as the *closed-loop transfer function* of the system. The numerator is the forward (open-loop) gain from r to y, and the denominator is one plus the gain in going around the feedback loop, the so-called loop gain. If $|P(s)C(s)| \gg 1$, i.e., it has a large norm with each value of s, and if $|F(s)| \approx 1$, then *Y(s)* is approximately equal to *R(s)* and the output closely tracks the reference input.

PID FEEDBACK CONTROL

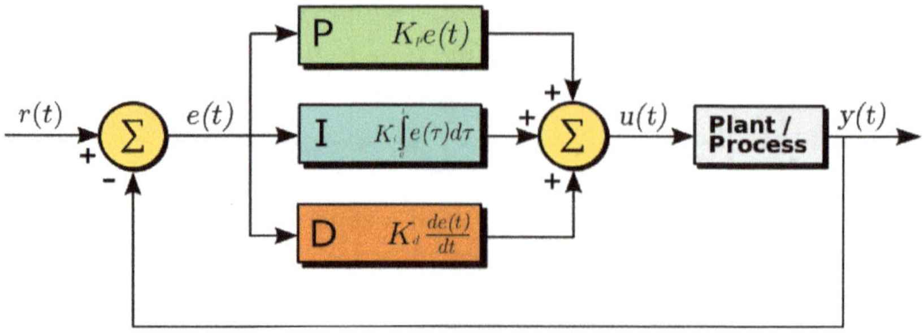

A block diagram of a PID controller in a feedback loop, r(t) is the desired process value or "set point", and y(t) is the measured process value.

A **proportional–integral–derivative controller** (**PID controller**) is a control loop feedback mechanism control technique widely used in control systems.

A PID controller continuously calculates an *error value* {\displaystyle e(t)} as the difference between a desired setpoint and a measured process variable and applies a correction based on proportional, integral, and derivative terms. *PID* is an initialism for *Proportional-Integral-Derivative*, referring to the three terms operating on the error signal to produce a control signal.

The theoretical understanding and application dates from the 1920s, and they are implemented in nearly all analogue control systems; originally in mechanical controllers, and then using discrete electronics and latterly in industrial process computers. The PID controller is probably the most-used feedback control design.

Referring to the equation below;

If *u(t)* is the control signal sent to the system, *y(t)* is the measured output and *r(t)* is the desired output, and tracking error $e(t) = r(t) - y(t)$, a PID controller has the general form

$$u(t) = K_P e(t) + K_I \int e(\tau) \mathrm{d}\tau + K_D \frac{\mathrm{de(t)}}{\mathrm{dt}}.$$

The desired closed loop dynamics is obtained by adjusting the three parameters K_P, K_I and K_D , often iteratively by "tuning" and without specific knowledge of a plant model. Stability can often be ensured using only the proportional term. The integral term permits the rejection of a step disturbance (often a striking specification in process control). The derivative term is used to provide damping or shaping of the response. PID controllers are the most well established class of control systems: however, they cannot be used in several more complicated cases, especially if MIMO systems are considered.

Applying Laplace transformation results in the transformed PID controller equation

$$u(s) = K_P e(s) + K_I \frac{1}{s} e(s) + K_D s e(s)$$

$$u(s) = \left(K_P + K_I \frac{1}{s} + K_D s \right) e(s)$$

with the PID controller transfer function

$$C(s) = \left(K_P + K_I \frac{1}{s} + K_D s\right)$$

There exists a nice example of the closed-loop system discussed above. If we take PID controller transfer function in series form

$$C(s) = K\left(1 + \frac{1}{sT_i}\right)(1 + sT_d)$$

1st order filter in feedback loop

$$F(s) = \frac{1}{1 + sT_f}$$

linear actuator with filtered input

$$P(s) = \frac{A}{1 + sT_p}, \text{ A = const}$$

and insert all this into expression for closed-loop transfer function H(s), then tuning is very easy: simply put

$$K = \frac{1}{A}, T_i = T_f, T_d = T_p$$

and get H(s) = 1 identically.

For practical PID controllers, a pure differentiator is neither physically realisable nor desirable due to amplification of noise and resonant modes in the system. Therefore, a phase-lead compensator type approach is used instead, or a differentiator with low-pass roll-off.

LINEAR AND NONLINEAR CONTROL THEORY

The field of control theory can be divided into two branches:

- *Linear control theory* – This applies to systems made of devices which obey the superposition principle, which means roughly that the output is proportional to the input. They are governed by linear differential equations. A major subclass is systems which in addition have parameters which do not change with time, called *linear time invariant* (LTI) systems. These systems are amenable to powerful frequency domain mathematical techniques of great generality, such as the Laplace transform, Fourier transform, Z transform, Bode plot, root locus, and Nyquist stability criterion. These lead to a description of the system using terms like bandwidth, frequency response, eigenvalues, gain, resonant frequencies, poles, and zeros, which give solutions for system response and design techniques for most systems of interest.
- *Nonlinear control theory* – This covers a wider class of systems that do not obey the superposition principle, and applies to more real-world systems, because all real control systems are nonlinear. These systems are often governed by nonlinear differential equations. The few mathematical techniques which

have been developed to handle them are more difficult and much less general, often applying only to narrow categories of systems. These include limit cycle theory, Poincaré maps, Lyapunov stability theorem, and describing functions. Nonlinear systems are often analyzed using numerical methods on computers, for example by simulating their operation using a simulation language. If only solutions near a stable point are of interest, nonlinear systems can often be linearized by approximating them by a linear system using perturbation theory, and linear techniques can be used.

ANALYSIS TECHNIQUES - FREQUENCY DOMAIN AND TIME DOMAIN

Mathematical techniques for analyzing and designing control systems fall into two different categories:

- *Frequency domain* – In this type the values of the state variables, the mathematical variables representing the system's input, output and feedback are represented as functions of frequency. The input signal and the system's transfer function are converted from time functions to functions of frequency by a transform such as the Fourier transform, Laplace transform, or Z transform. The advantage of this technique is that it results in a simplification of the mathematics; the *differential equations* that represent the system are replaced by *algebraic equations* in the frequency domain which are much simpler to solve. However, frequency domain techniques can only be used with linear systems, as mentioned above.
- *Time-domain state space representation* – In this type the values of the state variables are represented as functions of time. With this model the system being analyzed is represented by one or more differential equations. Since frequency domain techniques are limited to linear systems, time domain is widely used to analyze real-world nonlinear systems. Although these are more difficult to solve, modern computer simulation techniques such as simulation languages have made their analysis routine.

In contrast to the frequency domain analysis of the classical control theory, modern control theory utilizes the time-domain state space representation, a mathematical model of a physical system as a set of input, output and state variables related by first-order differential equations. To abstract from the number of inputs, outputs and states, the variables are expressed as vectors and the differential and algebraic equations are written in matrix form (the latter only being possible when the dynamical system is linear). The state space representation (also known as the "time-domain approach") provides a convenient and compact way to model and analyze systems with multiple inputs and outputs. With inputs and outputs, we would otherwise have to write down Laplace transforms to encode all the information about a system. Unlike the frequency domain approach, the use of the state-space representation is not limited to systems with linear components and zero initial conditions. "State space" refers to the space whose axes are the state variables. The state of the system can be represented as a point within that space.

SYSTEM INTERFACING - SISO & MIMO

Control systems can be divided into different categories depending on the number of inputs and outputs.

- Single-input single-output (SISO) – This is the simplest and most common type, in which one output is controlled by one control signal. Examples are the cruise control example above, or an audio system, in which the control input is the input audio signal and the output is the sound waves from the speaker.
- Multiple-input multiple-output (MIMO) – These are found in more complicated systems. For example, modern large telescopes such as the Keck and MMT have mirrors composed of many separate segments each controlled by an actuator. The shape of the entire mirror is constantly adjusted by a MIMO active optics control system using input from multiple sensors at the focal plane, to compensate for changes in the mirror shape due to thermal expansion, contraction, stresses as it is rotated and distortion of the wavefront due to turbulence in the atmosphere. Complicated systems such as nuclear reactors and human cells are simulated by computer as large MIMO control systems.

Topics in Control Theory

Stability

The *stability* of a general dynamical system with no input can be described with Lyapunov stability criteria.

- A linear system is called bounded-input bounded-output (BIBO) stable if its output will stay bounded for any bounded input.
- Stability for nonlinear systems that take an input is input-to-state stability (ISS), which combines Lyapunov stability and a notion similar to BIBO stability.

For simplicity, the following descriptions focus on continuous-time and discrete-time linear systems.

Mathematically, this means that for a causal linear system to be stable all of the poles of its transfer function must have negative-real values, i.e. the real part of each pole must be less than zero. Practically speaking, stability requires that the transfer function complex poles reside

- in the open left half of the complex plane for continuous time, when the Laplace transform is used to obtain the transfer function.
- inside the unit circle for discrete time, when the Z-transform is used.

The difference between the two cases is simply due to the traditional method of plotting continuous time versus discrete time transfer functions. The continuous Laplace transform is in Cartesian coordinates where the x axis is the real axis and the discrete Z-transform is in circular coordinates where the ρ axis axis is the real axis.

When the appropriate conditions above are satisfied a system is said to be asymptotically stable; the variables of an asymptotically stable control system always decrease from their initial value and do not show permanent oscillations. Permanent oscillations occur when a pole has a real part exactly equal to zero (in the continuous time case) or a modulus equal to one (in the discrete time case). If a simply stable system response neither decays nor grows over time, and has no oscillations, it is marginally stable; in this case the system transfer function has non-repeated poles at complex plane origin (i.e. their real and complex component is zero in the continuous time case). Oscillations are present when poles with real part equal to zero have an imaginary part not equal to zero.

If a system in question has an impulse response of

$$x[n] = 0.5^n u[n]$$

then the Z-transform (see this example), is given by

$$X(z) = \frac{1}{1 - 0.5z^{-1}}$$

which has a pole in z=5.0 (zero imaginary part). This system is BIBO (asymptotically) stable since the pole is *inside* the unit circle.

However, if the impulse response was

$$x[n] = 1.5^n u[n]$$

then the Z-transform is

$$X(z) = \frac{1}{1 - 1.5z^{-1}}$$

which has a pole at z=1.5 and is not BIBO stable since the pole has a modulus strictly greater than one.

Numerous tools exist for the analysis of the poles of a system. These include graphical systems like the root locus, Bode plots or the Nyquist plots.

Mechanical changes can make equipment (and control systems) more stable. Sailors add ballast to improve the stability of ships. Cruise ships use antiroll fins that extend transversely from the side of the ship for perhaps 30 feet (10 m) and are continuously rotated about their axes to develop forces that oppose the roll.

Controllability and Observability

Controllability and observability are main issues in the analysis of a system before deciding the best control strategy to be applied, or whether it is even possible to control or stabilize the system. Controllability is related to the possibility of forcing the system into a particular state by using an appropriate control signal. If a state is not controllable, then no signal will ever be able to control the state. If a state is not controllable, but its dynamics are stable, then the state is termed *stabilizable*. Observability instead is related to the possibility of *observing*, through output measurements, the state of a system. If a state is not observable, the controller will never be able to determine the behaviour of an unobservable state and hence cannot use it to stabilize the system. However, similar to the stabilizability condition above, if a state cannot be observed it might still be detectable.

From a geometrical point of view, looking at the states of each variable of the system to be controlled, every "bad" state of these variables must be controllable and observable to ensure a good behaviour in the closed-loop system. That is, if one of the eigenvalues of the system is not both controllable and observable, this part of the dynamics will remain untouched in the closed-loop system. If such an eigenvalue is not stable, the dynamics of this eigenvalue will be present in the closed-loop system which therefore will be unstable. Unobservable poles are not present in the transfer function realization of a state-space representation, which is why sometimes the latter is preferred in dynamical systems analysis.

Solutions to problems of uncontrollable or unobservable system include adding actuators and sensors.

Control Specification

Several different control strategies have been devised in the past years. These vary from extremely general ones (PID controller), to others devoted to very particular classes of systems (especially robotics or aircraft cruise control).

A control problem can have several specifications. Stability, of course, is always present. The controller must ensure that the closed-loop system is stable, regardless of the open-loop stability. A poor choice of controller can even worsen the stability of the open-loop system, which must normally be avoided. Sometimes it would be desired to obtain particular dynamics in the closed loop: i.e. that the poles have $Re[\lambda] < -\overline{\lambda}$, where $\overline{\lambda}$ is a fixed value strictly greater than zero, instead of simply asking that $Re[\lambda] < 0$.

Another typical specification is the rejection of a step disturbance; including an integrator in the open-loop chain (i.e. directly before the system under control) easily achieves this. Other classes of disturbances need different types of sub-systems to be included.

Other "classical" control theory specifications regard the time-response of the closed-loop system. These include the rise time (the time needed by the control system to reach the desired value after a perturbation), peak overshoot (the highest value reached by the response before reaching the desired value) and others (settling time, quarter-decay). Frequency domain specifications are usually related to robustness (see after).

Modern performance assessments use some variation of integrated tracking error (IAE,ISA,CQI).

Model identification and Robustness

A control system must always have some robustness property. A robust controller is such that its properties do not change much if applied to a system slightly different from the mathematical one used for its synthesis. This specification is important, as no real physical system truly behaves like the series of differential equations used to represent it mathematically. Typically a simpler mathematical model is chosen in order to simplify calculations, otherwise the true system dynamics can be so complicated that a complete model is impossible.

System Identification

The process of determining the equations that govern the model's dynamics is called system identification. This can be done off-line: for example, executing a series of measures from which to calculate an approximated mathematical model, typically its transfer function or matrix. Such identification from the output, however, cannot take account of unobservable dynamics. Sometimes the model is built directly starting from known physical equations, for example, in the case of a mass-spring-damper system we know that $m\ddot{x}(t) = -Kx(t) - \mathrm{B}\dot{x}(t)$. Even assuming that a "complete" model is used in designing the controller, all the parameters included in these equations (called "nominal parameters") are never known with absolute precision; the control system will have to behave correctly even when connected to physical system with true parameter values away from nominal.

Some advanced control techniques include an "on-line" identification process (see later). The parameters of the model are calculated ("identified") while the controller itself is running. In this way, if a drastic variation of the parameters ensues, for example, if the robot's arm releases a weight, the controller will adjust itself consequently in order to ensure the correct performance.

Analysis

Analysis of the robustness of a SISO (single input single output) control system can be performed in the frequency domain, considering the system's transfer function and using Nyquist and Bode diagrams. Topics include gain and phase margin and amplitude margin. For MIMO (multi input multi output) and, in general, more complicated control systems one must consider the theoretical results devised for each control technique (see next section). I.e., if particular robustness qualities are needed, the engineer must shift his attention to a control technique by including them in its properties.

Constraints

A particular robustness issue is the requirement for a control system to perform properly in the presence of input and state constraints. In the physical world every signal is limited. It could happen that a controller will send control signals that cannot be followed by the physical system, for example, trying to rotate a valve at excessive speed. This can produce undesired behavior of the closed-loop system, or even damage or break actuators or other subsystems. Specific control techniques are available to solve the problem: model predictive control (see later), and anti-wind up systems. The latter consists of an additional control block that ensures that the control signal never exceeds a given threshold.

SYSTEM CLASSIFICATIONS

Linear Systems Control

For MIMO systems, pole placement can be performed mathematically using a state space representation of the open-loop system and calculating a feedback matrix assigning poles in the desired positions. In complicated systems this can require computer-assisted calculation capabilities, and cannot always ensure robustness. Furthermore, all system states are not in general measured and so observers must be included and incorporated in pole placement design.

Nonlinear Systems Control

Processes in industries like robotics and the aerospace industry typically have strong nonlinear dynamics. In control theory it is sometimes possible to linearize such classes of systems and apply linear techniques, but in many cases it can be necessary to devise from scratch theories permitting control of nonlinear systems. These, e.g., feedback linearization, backstepping, sliding mode control, trajectory linearization control normally take advantage of results based on Lyapunov's theory. Differential geometry has been widely used as a tool for generalizing well-known linear control concepts to the non-linear case, as well as showing the subtleties that make it a more challenging problem. Control theory has also been used to decipher the neural mechanism that directs cognitive states.

Decentralized Systems Control

When the system is controlled by multiple controllers, the problem is one of decentralized control. Decentralization is helpful in many ways, for instance, it helps control systems to operate over a larger geographical area. The agents in decentralized control systems can interact using communication channels and coordinate their actions.

Deterministic and Stochastic Systems Control

A stochastic control problem is one in which the evolution of the state variables is subjected to random shocks from outside the system. A deterministic control problem is not subject to external random shocks.

Main control strategies

Every control system must guarantee first the stability of the closed-loop behavior. For linear systems, this can be obtained by directly placing the poles. Non-linear control systems use specific theories (normally based on Aleksandr Lyapunov's Theory) to ensure stability without regard to the inner dynamics of the system. The possibility to fulfill different specifications varies from the model considered and the control strategy chosen.

List of the main control techniques

- *Adaptive* control uses on-line identification of the process parameters, or modification of controller gains, thereby obtaining strong robustness properties. Adaptive controls were applied for the first time in the aerospace industry in the 1950s, and have found particular success in that field.
- *A hierarchical* control system is a type of control system in which a set of devices and governing software is arranged in a hierarchical tree. When the

links in the tree are implemented by a computer network, then that hierarchical control system is also a form of networked control system.

- *Intelligent* control uses various AI computing approaches like neural networks, Bayesian probability, fuzzy logic,machine learning, evolutionary computation and genetic algorithms to control a dynamic system.
- *Optimal* control is a particular control technique in which the control signal optimizes a certain "cost index": for example, in the case of a satellite, the jet thrusts needed to bring it to desired trajectory that consume the least amount of fuel. Two optimal control design methods have been widely used in industrial applications, as it has been shown they can guarantee closed-loop stability. These are Model Predictive Control (MPC) and linear-quadratic-Gaussian control (LQG). The first can more explicitly take into account constraints on the signals in the system, which is an important feature in many industrial processes. However, the "optimal control" structure in MPC is only a means to achieve such a result, as it does not optimize a true performance index of the closed-loop control system. Together with PID controllers, MPC systems are the most widely used control technique in process control.
- *Robust* control deals explicitly with uncertainty in its approach to controller design. Controllers designed using *robust control* methods tend to be able to cope with small differences between the true system and the nominal model used for design. The early methods of Bode and others were fairly robust; the state-space methods invented in the 1960s and 1970s were sometimes found to lack robustness. Examples of modern robust control techniques include H-infinity loop-shaping developed by Duncan McFarlane and Keith Glover of Cambridge University, United Kingdom and Sliding mode control (SMC) developed by Vadim Utkin. Robust methods aim to achieve robust performance and/or stability in the presence of small modeling errors.
- *Stochastic* control deals with control design with uncertainty in the model. In typical stochastic control problems, it is assumed that there exist random noise and disturbances in the model and the controller, and the control design must take into account these random deviations.
- *Energy-shaping* control view the plant and the controller as energy-transformation devices. The control strategy is formulated in terms of interconnection (in a power-preserving manner) in order to achieve a desired behavior.
- *Self-organized criticality* control may be defined as attempts to interfere in the processes by which the self-organized system dissipates energy.

HISTORY

Although control systems of various types date back to antiquity, a more formal analysis of the field began with a dynamics analysis of the centrifugal governor, conducted by the physicist James Clerk Maxwell in 1868, entitled *On Governors*.This described and analyzed the phenomenon of self-oscillation, in which lags in the system may lead to overcompensation and unstable behavior. This generated a flurry of interest in the topic, during which Maxwell's classmate, Edward John Routh, abstracted Maxwell's results for the general class of linear systems. Independently, Adolf Hurwitz analyzed system stability using differential equations in 1877, resulting in what is now known as the Routh–Hurwitz theorem.

Centrifugal governor in a Boulton & Watt engine of 1788.

A notable application of dynamic control was in the area of manned flight. The Wright brothers made their first successful test flights on December 17, 1903 and were distinguished by their ability to control their flights for substantial periods (more so than the ability to produce lift from an airfoil, which was known). Continuous, reliable control of the airplane was necessary for flights lasting longer than a few seconds.

By World War II, control theory was an important part of fire-control systems, guidance systems and electronics.

Sometimes, mechanical methods are used to improve the stability of systems. For example, ship stabilizers are fins mounted beneath the waterline and emerging laterally. In contemporary vessels, they may be gyroscopically controlled active fins, which have the capacity to change their angle of attack to counteract roll caused by wind or waves acting on the ship.

The Sidewinder missile uses small control surfaces placed at the rear of the missile with spinning disks on their outer surfaces and these are known as rollerons. Airflow over the disks spins them to a high speed. If the missile starts to roll, the gyroscopic force of the disks drives the control surface into the airflow, cancelling the motion. Thus, the Sidewinder team replaced a potentially complex control system with a simple mechanical solution.

The Space Race also depended on accurate spacecraft control, and control theory has also seen an increasing use in fields such as economics.

REFERENCES

1. Antunes, Ricardo; Gonzalez, Vicente (March 3, 2015). "A Production Model for Construction: A Theoretical Framework". Buildings. 5 *(1): 209–228.* doi:10.3390/buildings5010209.
2. "Feedback and control systems" - JJ Di Steffano, AR Stubberud, IJ Williams. Schaums outline series, McGraw-Hill 1967
3. Mayr, Otto (1970). The Origins of Feedback Control. Clinton, MA USA: The Colonial Press, Inc.
4. Mayr, Otto (1969). The Origins of Feedback Control. Clinton, MA USA: The Colonial Press, Inc.
5. Bennett, Stuart (1992). A history of control engineering, 1930-1955. IET. p. p. 48. ISBN 978-0-86341-299-8.
6. Ang, K.H., Chong, G.C.Y., and Li, Y. (2005). PID control system analysis, design, and technology, IEEE Trans Control Systems Tech, 13(4), pp.559–576.
7. trim point
8. Donald M Wiberg. State space & linear systems. Schaum's outline series. McGraw Hill. ISBN 0-07-070096-6.
9. Terrell, William (1999). "Some fundamental control theory I: Controllability, observability, and duality —AND— Some fundamental control Theory II: Feedback linearization of single input nonlinear systems". Amer. Math. Monthly. 106: 705–719 and 812–828. doi:10.2307/2589614.
10. Gu Shi; et al. (2015). "Controllability of structural brain networks (Article Number 8414)". Nature Communications (6). doi:10.1038/ncomms9414. Lay summary. Here we use tools from control and network theories to offer a mechanistic explanation for how the brain moves between cognitive states drawn from the network organization of white matter microstructure.
11. Liu, Jie; Wilson Wang; Farid Golnaraghi; Eric Kubica (2010). "A novel fuzzy framework for nonlinear system control". Fuzzy Sets and Systems. 161 *(21): 2746–2759. doi:10.1016/j.fss.2010.04.009.*
12. Maxwell, J.C. (1868). "On Governors". Proceedings of the Royal Society of London. 16: 270–283. doi:10.1098/rspl.1867.0055. JSTOR 112510.
13. Routh, E.J.; Fuller, A.T. (1975). Stability of motion. Taylor & Francis.

14. Routh, E.J. (1877). A Treatise on the Stability of a Given State of Motion, Particularly Steady Motion: Particularly Steady Motion. Macmillan and co.

15. Hurwitz, A. (1964). "On The Conditions Under Which An Equation Has Only Roots With Negative Real Parts". Selected Papers on Mathematical Trends in Control Theory.

16. Richard Bellman (1964) Control Theory, Scientific American 211(3):186–200

Dynamical System

Chapter Contents

A dynamical system is all about the evolution of something over time. To create a dynamical system we simply need to decide (1) what is the "something" that will evolve over time and (2) what is the rule that specifies how that something evolves with time. In this way, a dynamical system is simply a model describing the temporal evolution of a system.

The State Space

The first step in creating a dynamical system is to pin down what is the special "something" that we want to evolve with time. To do this, we need to come up with a set of variables that give a complete description of the system at any particular time.

By "complete description," we don't necessarily mean that the variables will completely describe a real life system we may be trying to model. But, the variables must completely describe the state of the mathematical system. In a dynamical system, if we know the values of these variables at a particular time, we know everything about the state of the system at that time. To model some real life system, the modeler must clearly make a choice of what variables will form the complete description for the mathematical model.

The variables that completely describe the state of the dynamical system are called the state variables. The set of all the possible values of the state variables is the state space.

The state space can be discrete, consisting of isolated points, such as if the state variables could only take on integer values. It could be continuous, consisting of a smooth set of points, such as if the state variables could take on any real value. In the case where the state space is continuous and finite-dimensional, it is often called the phase space, and the number of state variables is the dimension of the dynamical system. The state space can also be infinite-dimensional.

INTRODUCTION

In mathematics, a dynamical system is a system in which a function describes the time dependence of a point in a geometrical space. Examples include the mathematical models that describe the swinging of a clock pendulum, the flow of water in a pipe, and the number of fish each springtime in a lake.

The Lorenz attractor arises in the study of the Lorenz Oscillator, a dynamical system.

At any given time, a dynamical system has a state given by a set of real numbers (a vector) that can be represented by a point in an appropriate state space (a geometrical manifold). The evolution rule of the dynamical system is a function that describes what future states follow from the current state. Often the function is deterministic; in other words, for a given time interval only one future state follows from the current state;however, some systems are stochastic, in that random events also affect the evolution of the state variables.

OVERVIEW

The concept of a dynamical system has its origins in Newtonian mechanics. There, as in other natural sciences and engineering disciplines, the evolution rule of dynamical systems is an implicit relation that gives the state of the system for only a short time into the future. (The relation is either a differential equation, difference equation or other time scale.) To determine the state for all future times requires iterating the relation many times—each advancing time a small step. The iteration procedure is referred to as solving the system or integrating the system. If the system can be solved, given an initial point it is possible to determine all its future positions, a collection of points known as a trajectory or orbit.

Before the advent of computers, finding an orbit required sophisticated mathematical techniques and could be accomplished only for a small class of dynamical systems. Numerical methods implemented on electronic computing machines have simplified the task of determining the orbits of a dynamical system.

For simple dynamical systems, knowing the trajectory is often sufficient, but most dynamical systems are too complicated to be understood in terms of individual trajectories. The difficulties arise because:

- The systems studied may only be known approximately—the parameters of the system may not be known precisely or terms may be missing from the equations. The approximations used bring into question the validity or relevance of numerical solutions. To address these questions several notions of stability have been introduced in the study of dynamical systems, such as Lyapunov stability or structural stability. The stability of the dynamical system implies that there is a class of models or initial conditions for which the trajectories would be equivalent. The operation for comparing orbits to establish their equivalence changes with the different notions of stability.
- The type of trajectory may be more important than one particular trajectory. Some trajectories may be periodic, whereas others may wander through many different states of the system. Applications often require enumerating these classes or maintaining the system within one class. Classifying all possible trajectories has led to the qualitative study of dynamical systems, that is, properties that do not change under coordinate changes. Linear dynamical systems and systems that have two numbers describing a state are examples of dynamical systems where the possible classes of orbits are understood.
- The behavior of trajectories as a function of a parameter may be what is needed for an application. As a parameter is varied, the dynamical systems may have bifurcation points where the qualitative behavior of the dynamical system changes. For example, it may go from having only periodic motions to apparently erratic behavior, as in the transition to turbulence of a fluid.
- The trajectories of the system may appear erratic, as if random. In these cases it may be necessary to compute averages using one very long trajectory or many different trajectories. The averages are well defined for ergodic systems and a more detailed understanding has been worked out for hyperbolic systems. Understanding the probabilistic aspects of dynamical systems has helped establish the foundations of statistical mechanics and of chaos.

HISTORY

Many people regard Henri Poincaré as the founder of dynamical systems. Poincaré published two now classical monographs, "New Methods of Celestial Mechanics" (1892–1899) and "Lectures on Celestial Mechanics" (1905–1910). In them, he successfully applied the results of their research to the problem of the motion of three bodies and studied in detail the behavior of solutions (frequency, stability, asymptotic, and so on). These papers included the Poincaré recurrence theorem, which states that certain systems will, after a sufficiently long but finite time, return to a state very close to the initial state.

Aleksandr Lyapunov developed many important approximation methods. His methods, which he developed in 1899, make it possible to define the stability of sets of ordinary differential equations. He created the modern theory of the stability of a dynamic system.

In 1913, George David Birkhoff proved Poincaré's "Last Geometric Theorem", a special case of the three-body problem, a result that made him world-famous. In 1927, he published his Dynamical SystemsBirkhoff's most durable result has been his 1931 discovery of what is now called the ergodic theorem. Combining insights from physics on the ergodic hypothesis with measure theory, this theorem solved, at least in principle, a fundamental problem of statistical mechanics. The ergodic theorem has also had repercussions for dynamics.

Stephen Smale made significant advances as well. His first contribution is the Smale horseshoe that jumpstarted significant research in dynamical systems. He also outlined a research program carried out by many others.

Oleksandr Mykolaiovych Sharkovsky developed Sharkovsky's theorem on the periods of discrete dynamical systems in 1964. One of the implications of the theorem is that if a discrete dynamical system on the real line has a periodic point of period 3, then it must have periodic points of every other period.

BASIC DEFINITIONS

A dynamical system is a manifold M called the phase (or state) space endowed with a family of smooth evolution functions Φ^tthat for any element of $t \in T$, the time, map a point of the phase space back into the phase space. The notion of smoothness changes with applications and the type of manifold. There are several choices for the set T. When T is taken to be the reals, the dynamical system is called a flow; and if T is restricted to the non-negative reals, then the dynamical system is a semi-flow. When T is taken to be the integers, it is a cascade or a map; and the restriction to the non-negative integers is a semi-cascade.

EXAMPLES

The evolution function Φ^t is often the solution of a differential equation of motion

$$\dot{x} = v(x).$$

The equation gives the time derivative, represented by the dot, of a trajectory x(t) on the phase space starting at some point x_0. The vector field v(x) is a smooth function that at every point of the phase space M provides the velocity vector of the dynamical system at that point. (These vectors are not vectors in the phase space M, but in the tangent space T_xM of the point x.) Given a smooth Φ^t, an autonomous vector field can be derived from it. There is no need for higher order derivatives in the equation, nor for time dependence in v(x) because these can be eliminated by considering systems of higher dimensions. Other types of differential equations can be used to define the evolution rule:

$$G(x, \dot{x}) = 0$$

is an example of an equation that arises from the modeling of mechanical systems with complicated constraints.

The differential equations determining the evolution function Φ^t are often ordinary differential equations: in this case the phase space M is a finite dimensional manifold. Many of the concepts in dynamical systems can be extended to infinite-dimensional manifolds—those that are locally Banach spaces—in which case the differential equations are partial differential equations. In the late 20th century the dynamical system perspective to partial differential equations started gaining popularity.

Further examples

- Logistic map
- Complex quadratic polynomial
- Dyadic transformation
- Tent map
- Double pendulum
- Arnold's cat map
- Horseshoe map
- Baker's map is an example of a chaotic piecewise linear map
- Billiards and outer billiards
- Hénon map
- Lorenz system
- Circle map
- Rössler map
- Kaplan-Yorke map
- List of chaotic maps
- Swinging Atwood's machine
- Quadratic map simulation system
- Bouncing ball dynamics

LINEAR DYNAMICAL SYSTEMS

Linear dynamical systems can be solved in terms of simple functions and the behavior of all orbits classified. In a linear system the phase space is the N-dimensional Euclidean space, so any point in phase space can be represented by a vector with N numbers. The analysis of linear systems is possible because they satisfy a superposition principle:

if u(t) and w(t) satisfy the differential equation for the vector field (but not necessarily the initial condition), then so will u(t) + w(t).

Flows

For a flow, the vector field Φ(x) is an affine function of the position in the phase space, that is,

$$\dot{x} = \phi(x) = Ax + b,$$

with A a matrix, b a vector of numbers and x the position vector. The solution to this system can be found by using the superposition principle (linearity). The case $b \neq 0$ with $A = 0$ is just a straight line in the direction of b:

$$\Phi^t(x_1) = x_1 + bt.$$

When b is zero and $A \neq 0$ the origin is an equilibrium (or singular) point of the flow, that is, if $x_0 = 0$, then the orbit remains there. For other initial conditions, the equation of motion is given by the exponential of a matrix: for an initial point x_0,

$$\Phi^t(x_0) = e^{tA}x_0.$$

When b = 0, the eigenvalues of A determine the structure of the phase space. From the eigenvalues and the eigenvectors of A it is possible to determine if an initial point will converge or diverge to the equilibrium point at the origin.

The distance between two different initial conditions in the case $A \neq 0$ will change exponentially in most cases, either converging exponentially fast towards a point, or diverging exponentially fast. Linear systems display sensitive dependence on initial conditions in the case of divergence. For nonlinear systems this is one of the (necessary but not sufficient) conditions for chaotic behavior.

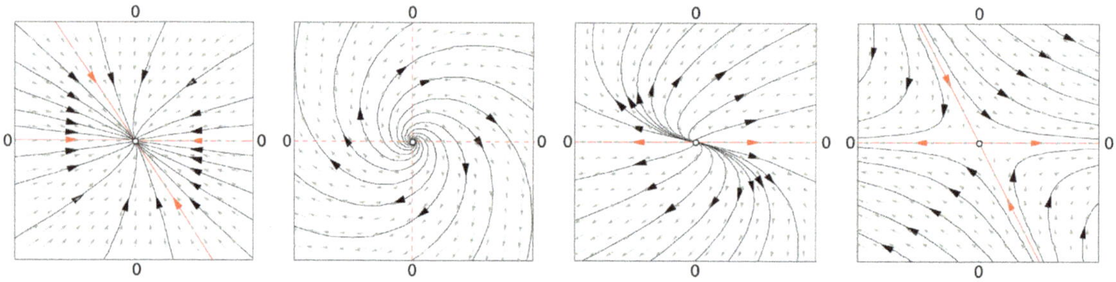

Linear vector fields and a few trajectories.

Maps

A discrete-time, affine dynamical system has the form of a matrix difference equation:

$$x_{n+1} = Ax_n + b,$$

with A a matrix and b a vector. As in the continuous case, the change of coordinates $x \rightarrow x + (1 - A)^{-1}b$ removes the term b from the equation. In the new coordinate system, the origin is a fixed point of the map and the solutions are of the linear system $A^n x_0$. The solutions for the map are no longer curves, but points that hop in the phase space. The orbits are organized in curves, or fibers, which are collections of points that map into themselves under the action of the map.

As in the continuous case, the eigenvalues and eigenvectors of A determine the structure of phase space. For example, if u_1 is an eigenvector of A, with a real eigenvalue smaller than one, then the straight lines given by the points along $\alpha\ u_1$, with $\alpha \in \mathbf{R}$, is an invariant curve of the map. Points in this straight line run into the fixed point.

There are also many other discrete dynamical systems.

LOCAL DYNAMICS

The qualitative properties of dynamical systems do not change under a smooth change of coordinates (this is sometimes taken as a definition of qualitative): a singular point of the vector field (a point where $v(x) = 0$) will remain a singular point under smooth transformations; a periodic orbit is a loop in phase space and smooth deformations of the phase space cannot alter it being a loop. It is in the neighborhood of singular points and periodic orbits that the structure of a phase space of a dynamical system can be well understood. In the qualitative study of dynamical systems, the approach is to show that there is a change of coordinates (usually unspecified, but computable) that makes the dynamical system as simple as possible.

Rectification

A flow in most small patches of the phase space can be made very simple. If y is a point where the vector field $v(y) \neq 0$, then there is a change of coordinates for a region around y where the vector field becomes a series of parallel vectors of the same magnitude. This is known as the rectification theorem.

The rectification theorem says that away from singular points the dynamics of a point in a small patch is a straight line. The patch can sometimes be enlarged by stitching several patches together, and when this works out in the whole phase space M the dynamical system is integrable. In most cases the patch cannot be extended to the entire phase space. There may be singular points in the vector field (where $v(x) = 0$); or the patches may become smaller and smaller as some point is approached. The more subtle reason is a global constraint, where the trajectory starts out in a patch, and after visiting a series of other patches comes back to the original one. If the next time the orbit loops around phase space in a different way, then it is impossible to rectify the vector field in the whole series of patches.

Near periodic orbits

In general, in the neighborhood of a periodic orbit the rectification theorem cannot be used. Poincaré developed an approach that transforms the analysis near a periodic orbit to the analysis of a map. Pick a point x_0 in the orbit γ and consider the points in phase space in that neighborhood that are perpendicular to $v(x_0)$. These points are a Poincaré section $S(\gamma, x_0)$, of the orbit. The flow now defines a map, the Poincaré map $F : S \rightarrow S$, for points starting in S and returning to S. Not all these points will take the same amount of time to come back, but the times will be close to the time it takes x_0.

The intersection of the periodic orbit with the Poincaré section is a fixed point of the Poincaré map F. By a translation, the point can be assumed to be at $x = 0$. The Taylor series of the map is $F(x) = J \cdot x + O(x^2)$, so a change of coordinates h can only be expected to simplify F to its linear part

$$h^{-1} \circ F \circ h(x) = J \cdot x.$$

This is known as the conjugation equation. Finding conditions for this equation to hold has been one of the major tasks of research in dynamical systems. Poincaré first approached it assuming all functions to be analytic and in the process discovered the non-resonant condition. If $\lambda_1, ..., \lambda_v$ are the eigenvalues of J they will be resonant if one eigenvalue is an integer linear combination of two or more of the others. As terms of the form $\lambda_i - \sum$ (multiples of other eigenvalues) occurs in the denominator of the terms for the function h, the non-resonant condition is also known as the small divisor problem.

Conjugation Results

The results on the existence of a solution to the conjugation equation depend on the eigenvalues of J and the degree of smoothness required from h. As J does not need to have any special symmetries, its eigenvalues will typically be complex numbers. When the eigenvalues of J are not in the unit circle, the dynamics near the fixed point x_0 of F is called hyperbolic and when the eigenvalues are on the unit circle and complex, the dynamics is called elliptic.

In the hyperbolic case, the Hartman–Grobman theorem gives the conditions for the existence of a continuous function that maps the neighborhood of the fixed point of the map to the linear map $J \cdot x$. The hyperbolic case is also structurally stable. Small changes in the vector field will only produce small changes in the Poincaré map and these small changes will reflect in small changes in the position of the eigenvalues of J in the complex plane, implying that the map is still hyperbolic.

The Kolmogorov–Arnold–Moser (KAM) theorem gives the behavior near an elliptic point.

BIFURCATION THEORY

When the evolution map Φ^t (or the vector field it is derived from) depends on a parameter μ, the structure of the phase space will also depend on this parameter. Small changes may produce no qualitative changes in the phase space until a special value μ_0 is reached. At this point the phase space changes qualitatively and the dynamical system is said to have gone through a bifurcation. Bifurcation theory considers a structure in phase space (typically a fixed point, a periodic orbit, or an invariant torus) and studies its behavior as a function of the parameter μ. At the bifurcation point the structure may change its stability, split into new structures, or merge with other structures. By using Taylor series approximations of the maps and an understanding of the differences that may be eliminated by a change of coordinates, it is possible to catalog the bifurcations of dynamical systems.

The bifurcations of a hyperbolic fixed point x_0 of a system family F_μ can be characterized by the eigenvalues of the first derivative of the system $DF_\mu(x_0)$ computed at the bifurcation point. For a map, the bifurcation will occur when there are eigenvalues of DF_μ on the unit circle. For a flow, it will occur when there are eigenvalues on the imaginary axis. For more information, see the main article on Bifurcation theory.

Some bifurcations can lead to very complicated structures in phase space. For example, the Ruelle–Takens scenario describes how a periodic orbit bifurcates into a torus and the torus into a strange attractor. In another example, Feigenbaum period-doubling describes how a stable periodic orbit goes through a series of period-doubling bifurcations.

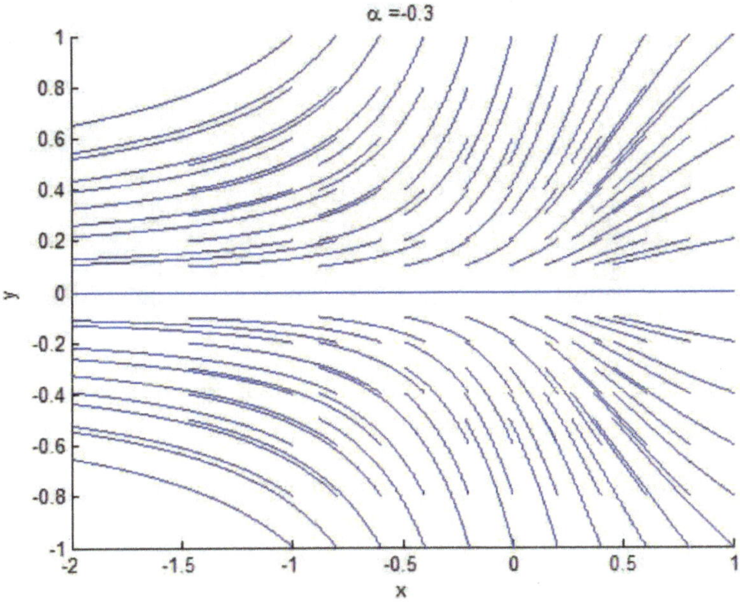

Phase portrait showing saddle-node bifurcation

Bifurcation theory is the mathematical study of changes in the qualitative or topological structure of a given family, such as the integral curves of a family of vector fields, and the solutions of a family of differential equations. Most commonly applied to the mathematical study of dynamical systems, a bifurcation occurs when a small smooth change made to the parameter values (the bifurcation parameters) of a system causes a sudden 'qualitative' or topological change in its behaviour.Bifurcations occur in both continuous systems (described by ODEs, DDEs or PDEs) and discrete systems (described by maps). The name "bifurcation" was first introduced by Henri Poincaré in 1885 in the first paper in mathematics showing such a behavior. Henri Poincaré also later named various types of stationary points and classified them.

Bifurcation Types

It is useful to divide bifurcations into two principal classes:

- Local bifurcations, which can be analysed entirely through changes in the local stability properties of equilibria, periodic orbits or other invariant sets as parameters cross through critical thresholds; and
- Global bifurcations, which often occur when larger invariant sets of the system 'collide' with each other, or with equilibria of the system. They cannot be detected purely by a stability analysis of the equilibria (fixed points).

Local Bifurcations

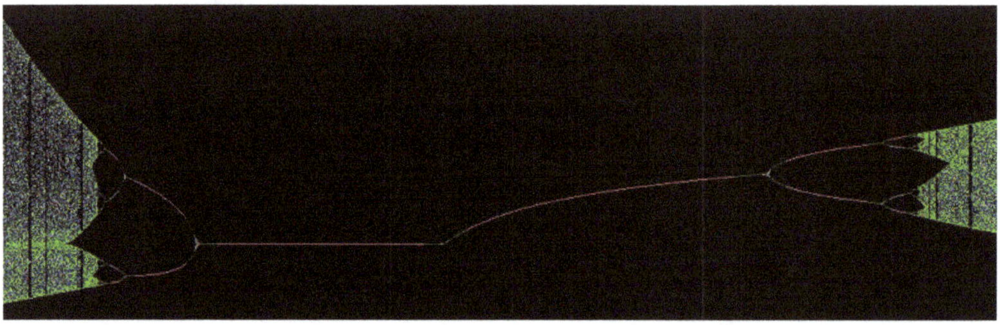

Period-halving bifurcations (L) leading to order, followed by period doubling bifurcations (R) leading to chaos.

A local bifurcation occurs when a parameter change causes the stability of an equilibrium (or fixed point) to change. In continuous systems, this corresponds to the real part of an eigenvalue of an equilibrium passing through zero. In discrete systems (those described by maps rather than ODEs), this corresponds to a fixed point having a Floquet multiplier with modulus equal to one. In both cases, the equilibrium is non-hyperbolic at the bifurcation point. The topological changes in the phase portrait of the system can be confined to arbitrarily small neighbourhoods of the bifurcating fixed points by moving the bifurcation parameter close to the bifurcation point (hence ‹local').

More technically, consider the continuous dynamical system described by the ODE

$$\dot{x} = f(x, \lambda) \quad f\colon \mathbb{R}^n \times \mathbb{R} \to \mathbb{R}^n.$$

A local bifurcation occurs at (x_0, λ_0) if the Jacobian matrix $\mathrm{d}f_{x_0,\lambda_0}$ has an eigenvalue with zero real part. If the eigenvalue is equal to zero, the bifurcation is a steady state bifurcation, but if the eigenvalue is non-zero but purely imaginary, this is a Hopf bifurcation.

For discrete dynamical systems, consider the system

$$x_{n+1} = f(x_n, \lambda)$$

Then a local bifurcation occurs at (x_0, λ_0) if the matrix $\mathbf{d}f_{x_0,\lambda_0}$ has an eigenvalue with modulus equal to one. If the eigenvalue is equal to one, the bifurcation is either a saddle-node (often called fold bifurcation in maps), transcritical or pitchfork bifurcation. If the eigenvalue is equal to −1, it is a period-doubling (or flip) bifurcation, and otherwise, it is a Hopf bifurcation.

Examples of local bifurcations include:

- Saddle-node (fold) bifurcation
- Transcritical bifurcation
- Pitchfork bifurcation
- Period-doubling (flip) bifurcation
- Hopf bifurcation
- Neimark–Sacker (secondary Hopf) bifurcation

Global Bifurcations

Global bifurcations occur when 'larger' invariant sets, such as periodic orbits, collide with equilibria. This causes changes in the topology of the trajectories in the phase space which cannot be confined to a small neighbourhood, as is the case with local bifurcations. In fact, the changes in topology extend out to an arbitrarily large distance (hence 'global').

Examples of global bifurcations include:

- Homoclinic bifurcation in which a limit cycle collides with a saddle point.
- Heteroclinic bifurcation in which a limit cycle collides with two or more saddle points.
- Infinite-period bifurcation in which a stable node and saddle point simultaneously occur on a limit cycle.
- Blue sky catastrophe in which a limit cycle collides with a nonhyperbolic cycle.

Global bifurcations can also involve more complicated sets such as chaotic attractors (e.g. crises).

Codimension of a Bifurcation

The codimension of a bifurcation is the number of parameters which must be varied for the bifurcation to occur. This corresponds to the codimension of the parameter set for which the bifurcation occurs within the full space of parameters. Saddle-node bifurcations and Hopf bifurcations are the only generic local bifurcations which are really codimension-one (the others all having higher codimension). However, transcritical and pitchfork bifurcations are also often thought of as codimension-one, because the normal forms can be written with only one parameter. An example of a well-studied codimension-two bifurcation is the Bogdanov–Takens bifurcation.

Applications in Semiclassical and Quantum Physics

Bifurcation theory has been applied to connect quantum systems to the dynamics of their classical analogues in atomic systems, molecular systems, and resonant tunneling diodes. Bifurcation theory has also been applied to the study of laser dynamics and a number of theoretical examples which are difficult to access experimentally such as the kicked topand coupled quantum wells. The dominant reason for the link between quantum systems and bifurcations in the classical equations of motion is that at bifurcations, the signature of classical orbits becomes large, as Martin Gutzwiller points out in his classic work on quantum chaos. Many kinds of bifurcations have been studied with regard to links between classical and quantum dynamics including saddle node bifurcations, Hopf bifurcations, umbilic bifurcations, period doubling bifurcations, reconnection bifurcations, tangent bifurcations, and cusp bifurcations.

ERGODIC SYSTEMS

In many dynamical systems, it is possible to choose the coordinates of the system so that the volume (really a ν-dimensional volume) in phase space is invariant. This happens for mechanical systems derived from Newton's laws as long as the coordinates are the position and the momentum and the volume is measured in units of (position) × (momentum). The flow takes points of a subset A into the points $\Phi^t(A)$ and invariance of the phase space means that

$$\mathrm{vol}(A) = \mathrm{vol}(\Phi^t(A)).$$

In the Hamiltonian formalism, given a coordinate it is possible to derive the appropriate (generalized) momentum such that the associated volume is preserved by the flow. The volume is said to be computed by the Liouville measure.

In a Hamiltonian system, not all possible configurations of position and momentum can be reached from an initial condition. Because of energy conservation, only the states with the same energy as the initial condition are accessible. The states with the same energy form an energy shell Ω, a sub-manifold of the phase space. The volume of the energy shell, computed using the Liouville measure, is preserved under evolution.

For systems where the volume is preserved by the flow, Poincaré discovered the recurrence theorem: Assume the phase space has a finite Liouville volume and let F be a phase space volume-preserving map and A a subset of the phase space. Then almost every point of A returns to A infinitely often. The Poincaré recurrence theorem was used by Zermelo to object to Boltzmann's derivation of the increase in entropy in a dynamical system of colliding atoms.

One of the questions raised by Boltzmann's work was the possible equality between time averages and space averages, what he called the ergodic hypothesis. The hypothesis states that the length of time a typical trajectory spends in a region A is vol(A)/vol(Ω).

The ergodic hypothesis turned out not to be the essential property needed for the development of statistical mechanics and a series of other ergodic-like properties were introduced to capture the relevant aspects of physical systems. Koopman approached the study of ergodic systems by the use of functional analysis. An observable a is a function that to each point of the phase space associates a number (say instantaneous pressure, or average height). The value of an observable can be computed at another time by using the evolution function φ^t. This introduces an operator U^t, the transfer operator,

$$(U^t a)(x) = a(\Phi^{-t}(x)).$$

By studying the spectral properties of the linear operator U it becomes possible to classify the ergodic properties of Φ^t. In using the Koopman approach of considering the action of the flow on an observable function, the finite-dimensional nonlinear problem involving Φ^t gets mapped into an infinite-dimensional linear problem involving U.

The Liouville measure restricted to the energy surface Ω is the basis for the averages computed in equilibrium statistical mechanics. An average in time along a trajectory is equivalent to an average in space computed with the Boltzmann factor exp(−βH). This idea has been generalized by Sinai, Bowen, and Ruelle (SRB) to a larger class of dynamical systems that includes dissipative systems. SRB measures replace the Boltzmann factor and they are defined on attractors of chaotic systems.

Nonlinear Dynamical Systems and Chaos

Simple nonlinear dynamical systems and even piecewise linear systems can exhibit a completely unpredictable behavior, which might seem to be random, despite the fact that they are fundamentally deterministic. This seemingly unpredictable behavior has been called chaos. Hyperbolic systems are precisely defined dynamical systems that exhibit the properties ascribed to chaotic systems. In hyperbolic systems the tangent space perpendicular to a trajectory can be well separated into two parts: one with the points that converge towards the orbit (the stable manifold) and another of the points that diverge from the orbit (the unstable manifold).

This branch of mathematics deals with the long-term qualitative behavior of dynamical systems. Here, the focus is not on finding precise solutions to the equations defining the dynamical system (which is often hopeless), but rather to answer questions like "Will the system settle down to a steady state in the long term, and if so, what are the possible attractors?" or "Does the long-term behavior of the system depend on its initial condition?"

Note that the chaotic behavior of complex systems is not the issue. Meteorology has been known for years to involve complex—even chaotic—behavior. Chaos theory has been so surprising because chaos can be found within almost trivial systems. The logistic map is only a second-degree polynomial; the horseshoe map is piecewise linear.

Geometrical Definition

A dynamical system is the tuple $\langle \mathcal{M}, f, \mathcal{T} \rangle$, with $\mathcal{M}$ a manifold (locally a Banach space or Euclidean space), $t \to f^t$ (with $t \in \mathcal{T}$) the domain for time (non-negative reals, the integers, ...) and f an evolution rule t → f^t (with $t \to f^t$ (with $t \in \mathcal{T}$) such that f^t is a diffeomorphism of the manifold to itself. So, f is a mapping of the time-domain

$\mathcal{T}$ into the space of diffeomorphisms of the manifold to itself. In other terms, f(t) is a diffeomorphism, for every time t in the domain $\mathcal{T}$.

Measure Theoretical Definition

A dynamical system may be defined formally, as a measure-preserving transformation of a sigma-algebra, the quadruplet (X, Σ, μ, τ). Here, X is a set, and Σ is a sigma-algebra on X, so that the pair (X, Σ) is a measurable space. μ is a finite measure on the sigma-algebra, so that the triplet (X, Σ, μ) is a probability space. A map τ: X → X is said to be Σ-measurable if and only if, for every σ ∈ Σ, one has $\sigma \in \Sigma$, one has $\tau^{-1}\sigma \in \Sigma$. A map τ is said to **preserve the measure** if and only if, for every σ ∈ Σ, one has $\mu(\tau^{-1}\sigma) = \mu(\sigma)$. Combining the above, a map τ is said to be a **measure-preserving transformation of *X***, if it is a map from X to itself, it is Σ-measurable, and is measure-preserving. The quadruple (X, Σ, μ, τ), for such a τ, is then defined to be a **dynamical system**.

The map τ embodies the time evolution of the dynamical system. Thus, for discrete dynamical systems the iterates $\tau^n = \tau \circ \tau \circ \cdots \circ \tau$ for integer n are studied. For continuous dynamical systems, the map τ is understood to be a finite time evolution map and the construction is more complicated.

EXAMPLES OF DYNAMICAL SYSTEMS

- Arnold's cat map
- Baker's map is an example of a chaotic piecewise linear map
- Circle map
- Double pendulum
- Billiards and Outer billiards
- Hénon map
- Horseshoe map
- Irrational rotation
- List of chaotic maps
- Logistic map
- Lorenz system
- Rossler map

MULTIDIMENSIONAL GENERALIZATION

Dynamical systems are defined over a single independent variable, usually thought of as time. A more general class of systems are defined over multiple independent variables and are therefore called multidimensional systems. Such systems are useful for modeling, for example, image processing.

ARNOLD'S CAT MAP

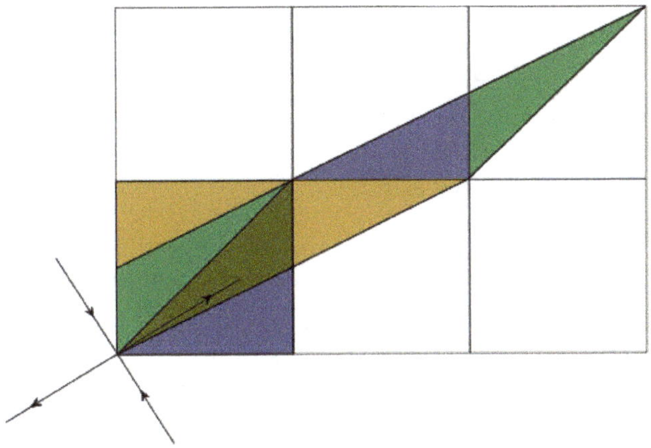

Picture showing how the linear map stretches the unit square and how its pieces are rearranged when the modulo operation is performed. The lines with the arrows show the direction of the contracting and expanding eigenspaces

In mathematics, Arnold's cat map is a chaotic map from the torus into itself, named after Vladimir Arnold, who demonstrated its effects in the 1960s using an image of a cat, hence the name.

Thinking of the torus $\mathbb{T}^2$ as the quotient space $\mathbb{R}^2/\mathbb{Z}^2$, Arnold's cat map is the transformation $\Gamma : \mathbb{T}^2 \to \mathbb{T}^2$ given by the formula

$$\Gamma : (x, y) \to (2x + y, x + y) \bmod 1.$$

Equivalently, in matrix notation, this is

$$\Gamma\left(\begin{bmatrix} x \\ y \end{bmatrix}\right) = \begin{bmatrix} 2 & 1 \\ 1 & 1 \end{bmatrix}\begin{bmatrix} x \\ y \end{bmatrix} \bmod 1 = \begin{bmatrix} 1 & 1 \\ 0 & 1 \end{bmatrix}\begin{bmatrix} 1 & 0 \\ 1 & 1 \end{bmatrix}\begin{bmatrix} x \\ y \end{bmatrix} \bmod 1.$$

That is, with a unit . equal to the width of the square image, the image is sheared one unit up, then two units to the right, and all that lies outside that unit square is shifted back by the unit until it's within the square.

Properties

- Γ is invertible because the matrix has determinant 1 and therefore its inverse has integer entries,
- Γ is area preserving,
- Γ has a unique hyperbolic fixed point (the vertices of the square). The linear transformation which defines the map is hyperbolic: its eigenvalues are irrational numbers, one greater and the other smaller than 1 (in absolute value), so they are associated respectively to an expanding and a contracting eigenspace which

are also the stable and unstable manifolds. The eigenspace are orthogonal because the matrix is symmetric. Since the eigenvectors have rationally independent components both the eigenspaces densely cover the torus. Arnold's cat map is a particularly well-known example of a hyperbolic toral automorphism, which is an automorphism of a torus given by a square unimodular matrix having no eigenvalues of absolute value 1.

- The set of the points with a periodic orbit is dense on the torus. Actually a point is preperiodic if and only if its coordinates are rational.
- Γ is topologically transitive (i.e. there is a point whose orbit is dense, this happens for example for any points on the expanding eigenspace)
- The number of points with period $|\lambda_1^n + \lambda_2^n - 2|$ (where λ_1 are the eigenvalues of the matrix). For example, the first few terms of this series are 1, 5, 16, 45, 121, 320, 841, 2205 .... (The same equation holds for any unimodular hyperbolic toral automorphism if the eigenvalues are replaced.)
- Γ is ergodic and mixing,
- Γ is an Anosov diffeomorphism and in particular it is structurally stable.

The Discrete Cat Map

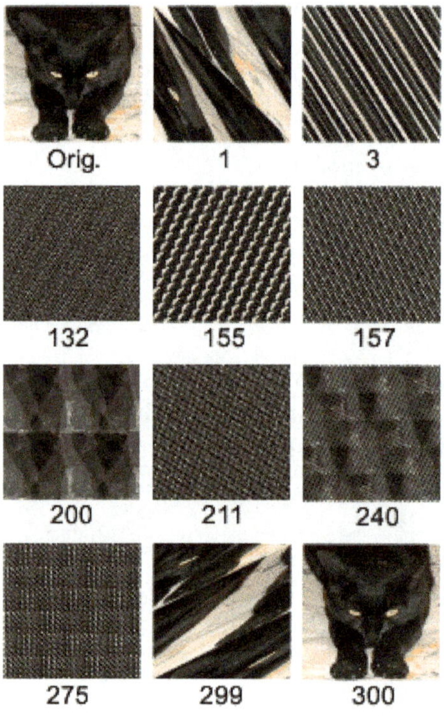

From order to chaos and back. Sample mapping on a picture of 150x150 pixels. The numbers shows the iteration step. After 300 iterations arriving at the original image

Sample mapping on a picture of a pair of cherries. The image is 74 pixels wide, and takes 114 iterations to be restored, although it appears upside-down at the halfway point.

It is possible to define a discrete analogue of the cat map. One of this map's features is that image being apparently randomized by the transformation but returning to its original state after a number of steps. As can be seen in the adjacent picture, the original image of the cat is sheared and then wrapped around in the first iteration of the transformation. After some iterations, the resulting image appears rather random or disordered, yet after further iterations the image appears to have further order—ghost-like images of the cat, multiple smaller copies arranged in a repeating structure and even upside-down copies of the original image—and ultimately returns to the original image.

The discrete cat map describes the phase space flow corresponding to the discrete dynamics of a bead hopping from site q_t ($0 \leq q_t < N$) to site q_{t+1} on a circular ring with circumference N, according to the second order equation:

$$q_{t+1} - 3q_t + q_{t-1} = 0 \mod N$$

Defining the momentum variable $p_t = q_t - q_{t-1}$, the above second order dynamics can be re-written as a mapping of the square $0 \leq q, p < N$ (the phase space of the discrete dynamical system) onto itself:

$$q_{t+1} = 2q_t + p_t \mod N$$
$$p_{t+1} = q_t + p_t \mod N$$

This Arnold cat mapping shows mixing behavior typical for chaotic systems. However, since the transformation has a determinant equal to unity, it is area-preserving and therefore invertible the inverse transformation being:

$$q_{t+1} = 2q_t + p_t \mod N$$
$$p_{t+1} = q_t + p_t \mod N$$

For real variables q and p, it is common to set N = 1. In that case a mapping of the unit square with periodic boundary conditions onto itself results.

When N is set to an integer value, the position and momentum variables can be restricted to integers and the mapping becomes a mapping of a toroidial square grid of points onto itself. Such an integer cat map is commonly used to demonstrate mixing behavior with Poincaré recurrence utilising digital images. The number of iterations needed to restore the image can be shown never to exceed 3N.

For an image, the relationship between iterations could be expressed as follows:

$$\begin{array}{llcl} n=0: & T^0(x,y) & = & \text{Input Image}(x,y) \\ n=1: & T^1(x,y) & = & T^0\left(\bmod(2x+y,N),\ \bmod(x+y,N)\right) \\ & & \vdots & \\ n=k: & T^k(x,y) & = & T^{k-1}\left(\bmod(2x+y,N),\ \bmod(x+y,N)\right) \\ & & \vdots & \\ n=m: & \text{Output Image}(x,y) & = & T^m(x,y) \end{array}$$

BAKER'S MAP

Example of a measure that is invariant under the action of the (unrotated) baker's map.

In dynamical systems theory, the **baker's map** is a chaotic map from the unit square into itself. It is named after a kneading operation that bakers apply to dough: the dough is cut in half, and the two halves are stacked on one another, and compressed.

The baker's map can be understood as the bilateral shift operator of a bi-infinite two-state lattice model. The baker's map is topologically conjugate to the horseshoe map. In physics, a chain of coupled baker's maps can be used to model deterministic diffusion.

As with many deterministic dynamical systems, the baker's map is studied by its action on the space of functions defined on the unit square. The baker's map defines an operator on the space of functions, known as the transfer operator of the map. The baker's map is an exactly solvable model of deterministic chaos, in that the eigenfunctions and eigenvalues of the transfer operator can be explicitly determined.

Formal Definition

There are two alternative definitions of the baker's map which are in common use. One definition folds over or rotates one of the sliced halves before joining it (similar to the horseshoe map) and the other does not.

The folded baker's map acts on the unit square as

$$S_{\text{baker-folded}}(x,y) = \begin{cases} (2x, y/2) & \text{for } 0 \le x < \frac{1}{2} \\ (2-2x, 1-y/2) & \text{for } \frac{1}{2} \le x < 1. \end{cases}$$

When the upper section is not folded over, the map may be written as

$$S_{\text{baker-unfolded}}(x,y) = \left(2x - \lfloor 2x \rfloor\,,\ \frac{y + \lfloor 2x \rfloor}{2}\right)$$

The folded baker's map is a two-dimensional analog of the tent map

$$S_{\text{tent}}(x) = \begin{cases} 2x & \text{for } 0 \le x < \frac{1}{2} \\ 2(1-x) & \text{for } \frac{1}{2} \le x < 1 \end{cases}$$

while the unfolded map is analogous to the Bernoulli map. Both maps are topologically conjugate. The Bernoulli map can be understood as the map that progressively lops digits off the dyadic expansion of x. Unlike the tent map, the baker's map is invertible.

Properties

The baker's map preserves the two-dimensional Lebesgue measure.

The map is strong mixing and it is topologically mixing.

The transfer operator U maps functions of the unit square to other functions on the unit square; it is given by

$$[Uf](x,y) = (f \circ S^{-1})(x,y).$$

The origin unit square is on top and the bottom shows the result as the square is swept from left to right.

The transfer operator is unitary on the Hilbert space of square-integrable functions on the unit square. The spectrum is continuous, and because the operator is unitary the eigenvalues lie on the unit circle. The transfer operator is not unitary on the space $\mathcal{P}_x \otimes L^2_y$ of functions polynomial in the first coordinate and square-integrable in the second. On this space, it has a discrete, non-unitary, decaying spectrum.

As a shift Operator

The baker's map can be understood as the two-sided shift operator on the symbolic dynamics of a one-dimensional lattice. Consider, for example, the bi-infinite string

$$\sigma = (\ldots, \sigma_{-2}, \sigma_{-1}, \sigma_0, \sigma_1, \sigma_2, \ldots)$$

where each position in the string may take one of the two binary values {\displaystyle \ sigma _{k}\in \{0,1\}}. The action of the shift operator on this string is

$$\tau(\ldots, \sigma_k, \sigma_{k+1}, \sigma_{k+2}, \ldots) = (\ldots, \sigma_{k-1}, \sigma_k, \sigma_{k+1}, \ldots)$$

that is, each lattice position is shifted over by one to the left. The bi-infinite string may be represented by two real numbers $0 \leq x, y \leq 1$ as

$$x(\sigma) = \sum_{k=0}^{\infty} \sigma_k 2^{-(k+1)}$$

and

$$y(\sigma) = \sum_{k=0}^{\infty} \sigma_{-k-1} 2^{-(k+1)}.$$

In this representation, the shift operator has the form

$$\tau(x, y) = \left(\frac{x + \lfloor 2y \rfloor}{2}, 2y - \lfloor 2y \rfloor \right)$$

which can be seen to be the inverse of the un-folded baker's map given above.

ARNOLD TONGUE

In mathematics, particularly in dynamical systems theory, an **Arnold tongue** of a finite-parameter family of circle maps, named after Vladimir Arnold, is a region in the space of parameters where the map has locally-constant rational rotation number. In other words, it is a level set of a rotation number with nonempty interior.

Circle map showing mode-locked regions or Arnold tongues in black. Ω varies from 0 to 1 along the x-axis, and K varies from 0 at the bottom to 4π at the top.

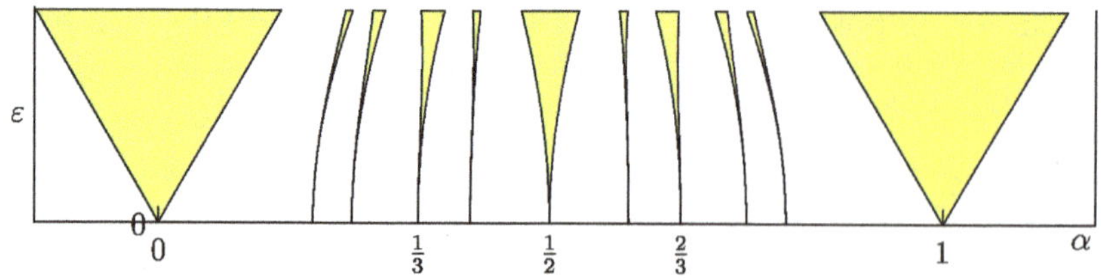

Some of the Arnold tongues for the standard circle map, $\varepsilon = \frac{K}{2\pi}$

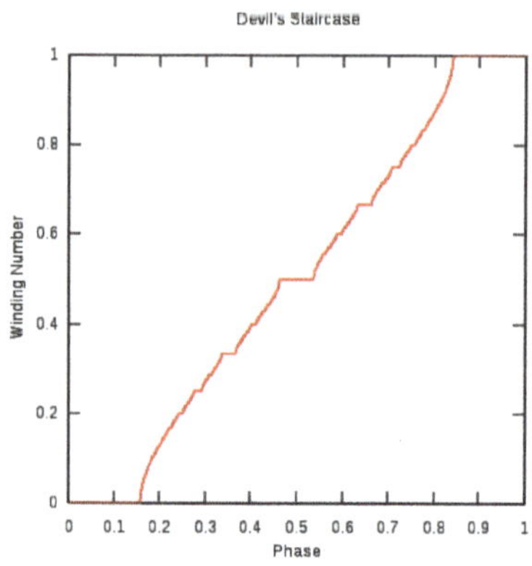

Rotation number as a function of Ω with K held constant at K = 1

Rotation number, with black corresponding to 0, green to 1/2 and red to 1. Ω varies from 0 to 1 along the x-axis, and K varies from 0 at the bottom to 2π at the top.

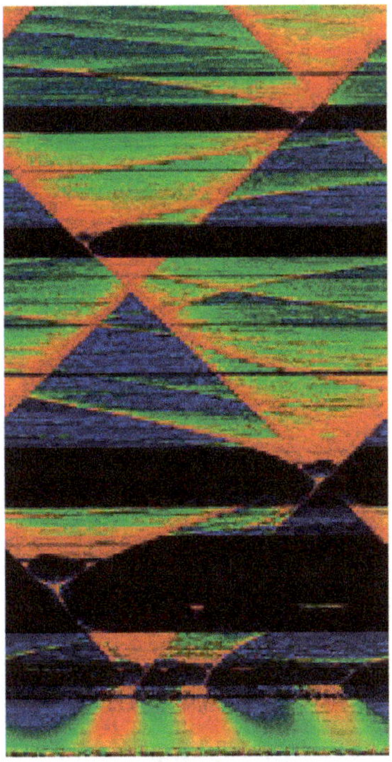

Bifurcation diagram for Ω held fixed at 1/3, and K running from 0 at bottom to 4π at top. Black regions correspond to Arnold tongues.

Standard Circle Map

Arnold tongues were first investigated for a family of dynamical systems on the circle first defined by Andrey Kolmogorov. Kolmogorov proposed this family as a simplified model for driven mechanical rotors (specifically, a free-spinning wheel weakly coupled by a spring to a motor). These circle map equations also describe a simplified model of the phase-locked loop in electronics. The map exhibits certain regions of its parameters where it is locked to the driving frequency (phase-locking or mode-locking in the language of electronic circuits). Among other applications, the circle map has been used to study the dynamical behaviour of a beating heart.

The **circle map** is given by iterating the map

$$\theta_{n+1} = \theta_n + \Omega - \frac{K}{2\pi}\sin(2\pi\theta_n).$$

where θ is to be interpreted as polar angle such that its value lies between 0 and 1.

It has two parameters, the coupling strength K and the driving phase Ω. As a model for phase-locked loops, Ω may be interpreted as a driving frequency. For K = 0 and Ω irrational, the map reduces to an irrational rotation.

Mode Locking

For small to intermediate values of K (that is, in the range of K = 0 to about K = 1), and certain values of Ω, the map exhibits a phenomenon called **mode locking** or **phase locking**. In a phase-locked region, the values θ_n advance essentially as a rational multiple of n, although they may do so chaotically on the small scale.

The limiting behavior in the mode-locked regions is given by the rotation number

$$\omega = \lim_{n \to \infty} \frac{\theta_n}{n}.$$

which is also sometimes referred to as the map winding number.

The phase-locked regions, or Arnold tongues, are illustrated in yellow in the figure above. Each such V-shaped region touches down to a rational value $\Omega = p/q$ in the limit of $K \to 0$. The values of (K,Ω) in one of these regions will all result in a motion such that the rotation number $\omega = p/q$. For example, all values of (K,Ω) in the large V-shaped region in the bottom-center of the figure correspond to a rotation number of $\omega = 1/2$. One reason the term "locking" is used is that the individual values θ_n can be perturbed by rather large random disturbances (up to the width of the tongue, for a given value of K), without disturbing the limiting rotation number. That is, the sequence stays "locked on" to the signal, despite the addition of significant noise to the series θ_n. This ability to "lock on" in the presence of noise is central to the utility of the phase-locked loop electronic circuit.

There is a mode-locked region for every rational number p/q. It is sometimes said that the circle map maps the rationals, a set of measure zero at K = 0, to a set of non-zero measure for $K \neq 0$. The largest tongues, ordered by size, occur at the Farey fractions. Fixing K and taking a cross-section through this image, so that ω is plotted as a function of Ω, gives the "Devil's staircase", a shape that is generically similar to the Cantor function.

The circle map also exhibits subharmonic routes to chaos, that is, period doubling of the form 3,6,12,24,....

Chirikov Standard Map

The Chirikov standard map is related to the circle map, having similar recurrence relations, which may be written as

$$\begin{aligned} \theta_{n+1} &= \theta_n + p_n + \frac{K}{2\pi}\sin(2\pi\theta_n) \\ p_{n+1} &= \theta_{n+1} - \theta_n \end{aligned}$$

with both iterates taken modulo 1. In essence, the standard map introduces a momentum p_n which is allowed to dynamically vary, rather than being forced fixed, as it is in the circle map. The standard map is studied in physics by means of the kicked rotor Hamiltonian.

Applications

A practical example of the application of Arnold's tongues to the synchronisation of a resonant tunneling diode oscillator is described in "Chaotic Dynamics in Resonant Tunneling Optoelectronic Voltage Controlled Oscillators", Romeira, B. et al , IEEE Photonics Technology Letter, Vol. 21, No. 24, 1819-1821, 2009, http://ieeexplore.ieee.org/stamp/stamp.jsp?tp=&arnumber=5325804&isnumber=5339268

DOUBLE PENDULUM

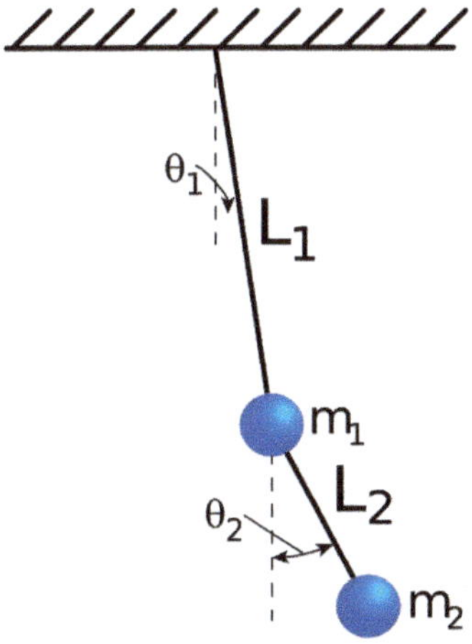

A double pendulum consists of two pendulums attached end to end.

In physics and mathematics, in the area of dynamical systems, a **double pendulum** is a pendulum with another pendulum attached to its end, and is a simple physical system that exhibits rich dynamic behavior with a strong sensitivity to initial conditions. The motion of a double pendulum is governed by a set of coupled ordinary differential equations and is chaotic.

Analysis and Interpretation

Several variants of the double pendulum may be considered; the two limbs may be of equal or unequal lengths and masses, they may be simple pendulums or compound pendulums (also called complex pendulums) and the motion may be in three dimensions or restricted to the vertical plane. In the following analysis, the limbs are taken to be identical compound pendulums of length l, and mass m the motion is restricted to two dimensions.

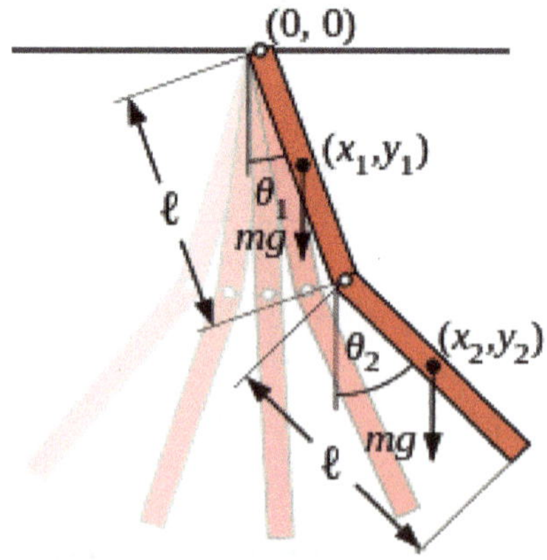

Double compound pendulum

In a compound pendulum, the mass is distributed along its length. If the mass is evenly distributed, then the center of mass of each limb is at its midpoint, and the limb has a moment of inertia of $I = \frac{1}{12}m\ell^2$ about that point.

It is convenient to use the angles between each limb and the vertical as the generalized coordinates defining the configuration of the system. These angles are denoted θ_1 and θ_2. The position of the center of mass of each rod may be written in terms of these two coordinates. If the origin of the Cartesian coordinate system is taken to be at the point of suspension of the first pendulum, then the center of mass of this pendulum is at:

$$x_1 = \frac{\ell}{2}\sin\theta_1,$$
$$y_1 = -\frac{\ell}{2}\cos\theta_1$$

and the center of mass of the second pendulum is at

$$x_2 = \ell\left(\sin\theta_1 + \frac{1}{2}\sin\theta_2\right),$$
$$y_2 = -\ell\left(\cos\theta_1 + \frac{1}{2}\cos\theta_2\right).$$

This is enough information to write out the Lagrangian.

Lagrangian

The Lagrangian is

$$\begin{aligned} L &= \text{Kinetic Energy} - \text{Potential Energy} \\ &= \frac{1}{2}m\left(v_1^2 + v_2^2\right) + \frac{1}{2}I\left(\dot{\theta}_1{}^2 + \dot{\theta}_2{}^2\right) - mg\left(y_1 + y_2\right) \\ &= \frac{1}{2}m\left(\dot{x}_1{}^2 + \dot{y}_1{}^2 + \dot{x}_2{}^2 + \dot{y}_2{}^2\right) + \frac{1}{2}I\left(\dot{\theta}_1{}^2 + \dot{\theta}_2{}^2\right) - mg\left(y_1 + y_2\right) \end{aligned}$$

The first term is the linear kinetic energy of the center of mass of the bodies and the second term is the rotational kinetic energy around the center of mass of each rod. The last term is the potential energy of the bodies in a uniform gravitational field. The dot-notation indicates the time derivative of the variable in question.

Substituting the coordinates above and rearranging the equation gives

$$L = \frac{1}{6} m\ell^2 \left[\dot{\theta}_2^{\ 2} + 4\dot{\theta}_1^{\ 2} + 3\dot{\theta}_1 \dot{\theta}_2 \cos(\theta_1 - \theta_2) \right] + \frac{1}{2} mg\ell \left(3\cos\theta_1 + \cos\theta_2 \right).$$

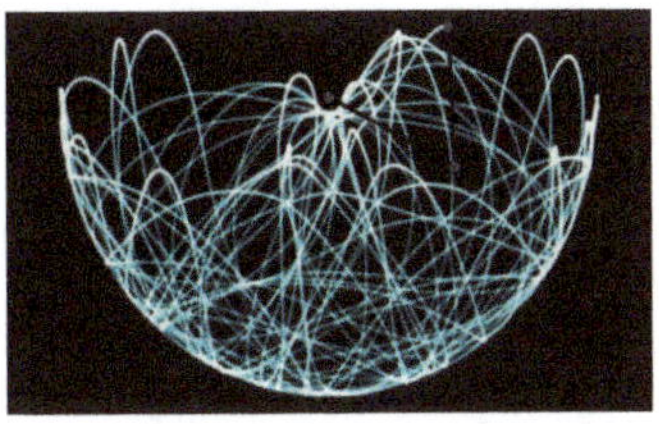

Motion of the double compound pendulum (from numerical integration of the equations of motion)

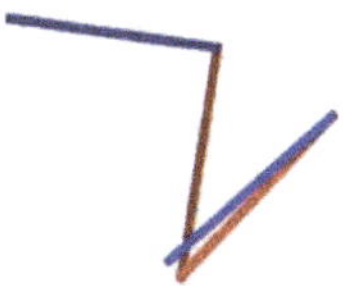

Trajectories of a double pendulum

Long exposure of double pendulum exhibiting chaotic motion (tracked with an LED)

There is only one conserved quantity (the energy), and no conserved momenta. The two momenta may be written as

$$p_{\theta_1} = \frac{\partial L}{\partial \dot{\theta}_1} = \frac{1}{6} m\ell^2 \left[8\dot{\theta}_1 + 3\dot{\theta}_2 \cos(\theta_1 - \theta_2) \right]$$

and

$$p_{\theta_2} = \frac{\partial L}{\partial \dot{\theta}_2} = \frac{1}{6} m\ell^2 \left[2\dot{\theta}_2 + 3\dot{\theta}_1 \cos(\theta_1 - \theta_2) \right].$$

These expressions may be inverted to get

$$\dot{\theta}_1 = \frac{6}{m\ell^2} \frac{2p_{\theta_1} - 3\cos(\theta_1 - \theta_2)p_{\theta_2}}{16 - 9\cos^2(\theta_1 - \theta_2)}$$

and

$$\dot{\theta}_2 = \frac{6}{m\ell^2} \frac{8p_{\theta_2} - 3\cos(\theta_1 - \theta_2)p_{\theta_1}}{16 - 9\cos^2(\theta_1 - \theta_2)}.$$

The remaining equations of motion are written as

$$\dot{p}_{\theta_1} = \frac{\partial L}{\partial \theta_1} = -\frac{1}{2} m\ell^2 \left[\dot{\theta}_1 \dot{\theta}_2 \sin(\theta_1 - \theta_2) + 3\frac{g}{\ell} \sin\theta_1 \right]$$

and

$$\dot{p}_{\theta_2} = \frac{\partial L}{\partial \theta_2} = -\frac{1}{2} m\ell^2 \left[-\dot{\theta}_1 \dot{\theta}_2 \sin(\theta_1 - \theta_2) + \frac{g}{\ell} \sin\theta_2 \right].$$

These last four equations are explicit formulae for the time evolution of the system given its current state. It is not possible to go further and integrate these equations analytically, to get formulae for θ_1 and θ_2 as functions of time. It is however possible to perform this integration numerically using the Runge Kutta method or similar techniques.

Chaotic Motion

Graph of the time for the pendulum to flip over as a function of initial conditions

The double pendulum undergoes chaotic motion, and shows a sensitive dependence on initial conditions. The image to the right shows the amount of elapsed time before the pendulum "flips over," as a function of initial conditions. Here, the initial value of θ_1 ranges along the x-direction, from −3 to 3. The initial value θ_2 ranges along the y-direction, from −3 to 3. (Presumably, this exposition is describing a stationary release with kinetic terms at zero.) The colour of each pixel indicates whether either pendulum flips within $10\sqrt{\ell/g}$ (green), within $100\sqrt{\ell/g}$ (red), $1000\sqrt{\ell/g}$ (purple) or $10000\sqrt{\ell/g}$ (blue). Initial conditions that don't lead to a flip within $10000\sqrt{\ell/g}$ are plotted white.

The boundary of the central white region is defined in part by energy conservation with the following curve:

$$3\cos\theta_1 + \cos\theta_2 = 2.$$

Within the region defined by this curve, that is if

$$3\cos\theta_1 + \cos\theta_2 > 2,$$

then it is energetically impossible for either pendulum to flip. Outside this region, the pendulum can flip, but it is a complex question to determine when it will flip. Similar behavior is observed for a double pendulum composed of two point masses rather than two rods with distributed mass.

The lack of a natural excitation frequency has led to the use of double pendulum systems in seismic resistance designs in buildings, where the building itself is the primary inverted pendulum, and a secondary mass is connected to complete the double pendulum.

HORSESHOE MAP

In the mathematics of chaos theory, a horseshoe map is any member of a class of chaotic maps of the square into itself. It is a core example in the study of dynamical systems. The map was introduced by Stephen Smale while studying the behavior of the orbits of the van der Pol oscillator. The action of the map is defined geometrically by squishing the square, then stretching the result into a long strip, and finally folding the strip into the shape of a horseshoe.

Most points eventually leave the square under the action of the map. They go to the side caps where they will, under iteration, converge to a fixed point in one of the caps. The points that remain in the square under repeated iteration form a fractal set and are part of the invariant set of the map.

The squishing, stretching and folding of the horseshoe map are typical of chaotic systems, but not necessary or even sufficient.

Mixing in a real ball of colored putty after consecutive iterations of Smale horseshoe map

In the horseshoe map, the squeezing and stretching are uniform. They compensate each other so that the area of the square does not change. The folding is done neatly, so that the orbits that remain forever in the square can be simply described.

For a horseshoe map:

- there are an infinite number of periodic orbits;
- periodic orbits of arbitrarily long period exist;
- the number of periodic orbits grows exponentially with the period; and
- close to any point of the fractal invariant set there is a point of a periodic orbit.

The Horseshoe Map

The horseshoe map f is a diffeomorphism defined from a region S of the plane into itself. The region S is a square capped by two semi-disks. The action of f is defined through the composition of three geometrically defined transformations. First the square is contracted along the vertical direction by a factor a < 1/2. The caps are contracted so as to remain semi-disks attached to the resulting rectangle. Contracting by a factor smaller than one half assures that there will be a gap between the branches of the horseshoe. Next the rectangle is stretched horizontally by a factor of 1/a; the caps remain unchanged. Finally the resulting strip is folded into a horseshoe-shape and placed back into S.

The interesting part of the dynamics is the image of the square into itself. Once that part is defined, the map can be extended to a diffeomorphism by defining its action on the caps. The caps are made to contract and eventually map inside one of the caps (the left one in the figure). The extension of f to the caps adds a fixed point to the non-wandering set of the map. To keep the class of horseshoe maps simple, the curved region of the horseshoe should not map back into the square.

The horseshoe map is one-to-one, which means that an inverse f^{-1} exists when restricted to the image of S under f.

By folding the contracted and stretched square in different ways, other types of horseshoe maps are possible.

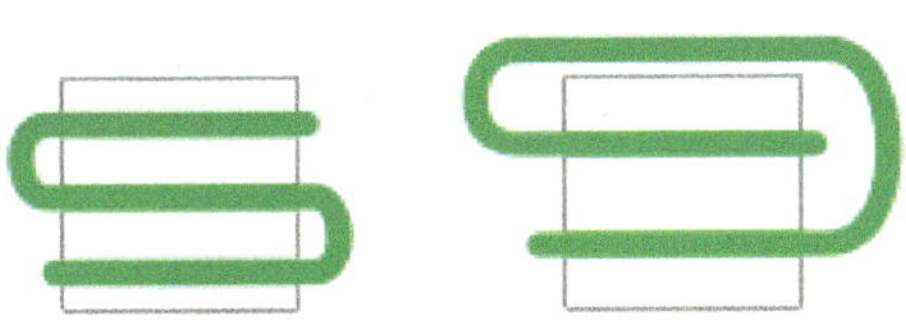

Variants of the horseshoe map

To ensure that the map remains one-to-one, the contracted square must not overlap itself. When the action on the square is extended to a diffeomorphism, the extension cannot always be done in the plane. For example, the map on the right needs to be extended to a diffeomorphism of the sphere by using a “cap” that wraps around the equator. The horseshoe map is an Axiom A diffeomorphism that serves as a model for the general behavior at a transverse homoclinic point, where the stable and unstable manifolds of a periodic point intersect.

Dynamics of the Map

The horseshoe map was designed to reproduce the chaotic dynamics of a flow in the neighborhood of a given periodic orbit. The neighborhood is chosen to be a small disk perpendicular to the orbit. As the system evolves, points in this disk remain close to the given periodic orbit, tracing out orbits that eventually intersect the disk once again. Other orbits diverge.

The behavior of all the orbits in the disk can be determined by considering what happens to the disk. The intersection of the disk with the given periodic orbit comes back to itself every period of the orbit and so do points in its neighborhood. When this neighborhood returns, its shape is transformed. Among the points back inside the disk are some points that will leave the disk neighborhood and others that will continue to return. The set of points that never leaves the neighborhood of the given periodic orbit form a fractal. A symbolic name can be given to all the orbits that remain in the neighborhood. The initial neighborhood disk can be divided into a small number of regions. Knowing the sequence in which the orbit visits these regions allows the orbit to be pinpointed exactly. The visitation sequence of the orbits provide a symbolic representation of the dynamics, known as symbolic dynamics.

Orbits

It is possible to describe the behavior of all initial conditions of the horseshoe map. An initial point $u_0 = (x, y)$ gets mapped into the point $u_1 = f(u_0)$. Its iterate is the point $u_2 = f(u_1) = f^2(u_0)$, and repeated iteration generates the orbit $u_0, u_1, u_2, \ldots$

Under repeated iteration of the horseshoe map, most orbits end up at the fixed point in the left cap. This is because the horseshoe maps the left cap into itself by an affine transformation that has exactly one fixed point. Any orbit that lands on the left cap never leaves it and converges to the fixed point in the left cap under iteration. Points in the right cap get mapped into the left cap on the next iteration, and most points in the square get mapped into the caps. Under iteration, most points will be part of orbits that converge to the fixed point in the left cap, but some points of the square never leave.

Iterating the Square

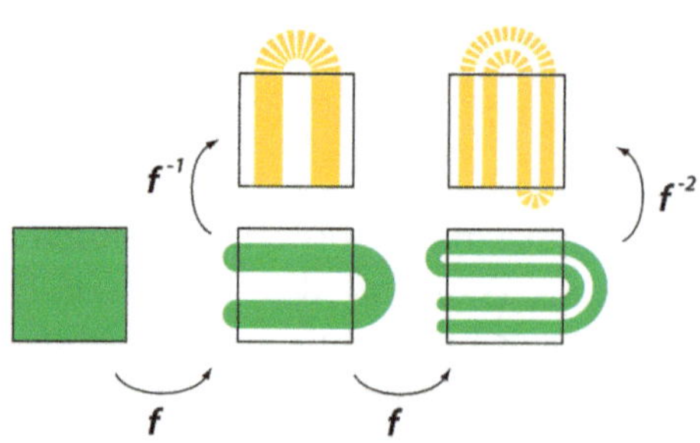

Pre-images of the square region

Under forward iterations of the horseshoe map, the original square gets mapped into a series of horizontal strips. The points in these horizontal strips come from vertical strips in the original square. Let S_0 be the original square, map it forward n times, and consider only the points that fall back into the square S_0, which is a set of horizontal stripes

$$H_n = f^n(S_0) \cap S_0.$$

The points in the horizontal stripes came from the vertical stripes

$$V_n = f^{-n}(H_n),$$

which are the horizontal strips H_n mapped backwards n times. That is, a point in V_n will, under n iterations of the horseshoe, end up in the set H_n of vertical strips.

Invariant set

Intersections that converge to the invariant set

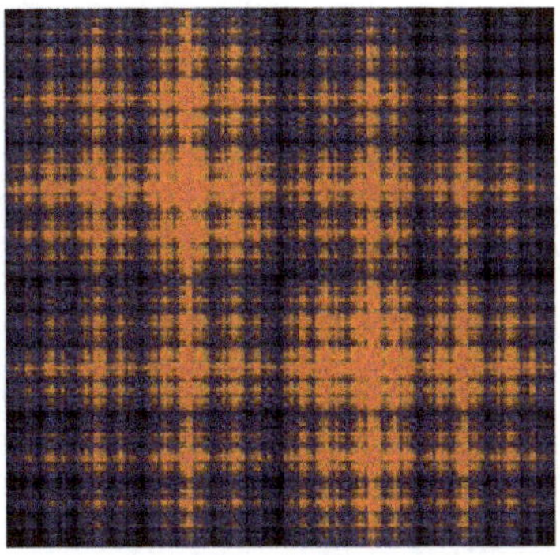

Example of an invariant measure

If a point is to remain indefinitely in the square, then it must belong to a set Λ that maps to itself. Whether this set is empty or not has to be determined. The vertical strips V_1 map into the horizontal strips H_1, but not all points of V_1 map back into V_1. Only the points in the intersection of V_1 and H_1 may belong to Λ, as can be checked by following points outside the intersection for one more iteration. The intersection of the horizontal and vertical stripes, $H_n \cap V_n$, are squares that in the limit $n \to \infty$ converge to the invariant set Λ. The structure of this set can be better understood by introducing a system of labels for all the intersections—a symbolic dynamics.

Symbolic Dynamics

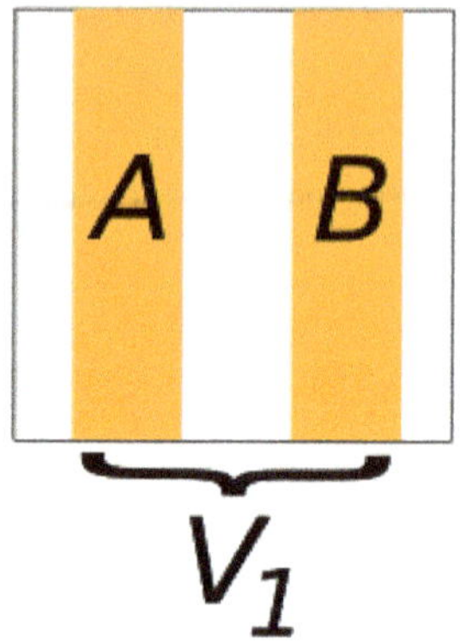

The basic domains of the horseshoe map

Since $H_n \cap V_n \subset V_1$, any point that is in Λ under iteration must land in the left vertical strip Aof V_1, or on the right vertical strip B. The lower horizontal strip of H_1 is the image of A and the upper horizontal strip is the image of B, so $H_1 = f(A) \cup f(B)$. The strips A and B can be used to label the four squares in the intersection of V_1 and H_1:

$$\Lambda_{A\bullet A} = f(A) \cap A$$
$$\Lambda_{A\bullet B} = f(A) \cap B$$
$$\Lambda_{B\bullet A} = f(B) \cap A$$
$$\Lambda_{B\bullet B} = f(B) \cap B$$

The set $\Lambda_{B\bullet A}$ consist of points from strip A that were in strip B in the previous iteration. A dot is used to separate the region the point of an orbit is in from the region the point came from.

The notation can be extended to higher iterates of the horseshoe map. The vertical strips can be named according to the sequence of visits to strip A or strip B. For example, the set ABB $\subset V_3$ consists of the points from A that will all land in B in one iteration and remain in B in the iteration after that:

$$ABB = \{x \in A \mid f(x) \in B \text{ and } f_2(x) \in B\}.$$

Working backwards from that trajectory determines a small region, the set ABB, within V_3.

The horizontal strips are named from their vertical strip pre-images. In this notation, the intersection of V_2 and H_2 consists of 16 squares, one of which is

$$\Lambda_{AB \bullet BB} = f_2(AB) \cap BB.$$

All the points in $\Lambda_{AB \bullet BB}$ are in B and will continue to be in B for at least one more iteration. Their previous trajectory before landing in BB was A followed by B.

Periodic Orbits

Any one of the intersections $\Lambda_{P \bullet F}$ of a horizontal strip with a vertical strip, where P and F are sequences of As and Bs, is an affine transformation of a small region in V_1. If P has k symbols in it, and if $f^{-k}(\Lambda_{P \bullet F})$ and $\Lambda_{P \bullet F}$ intersect, the region $\Lambda_{P \bullet F}$ will have a fixed point. This happens when the sequence P is the same as F. For example, $\Lambda_{ABAB \bullet ABAB} \subset V_4 \cap H_4$ has at least one fixed point. This point is also the same as the fixed point in $\Lambda_{AB \bullet AB}$. By including more and more ABs in the P and F part of the label of intersection, the area of the intersection can be made as small as needed. It converges to a point that is part of a periodic orbit of the horseshoe map. The periodic orbit can be labeled by the simplest sequence of As and Bs that labels one of the regions the periodic orbit visits.

For every sequence of As and Bs there is a periodic orbit.

OUTER BILLIARD

Outer billiards is a dynamical system based on a convex shape in the plane. Classically, this system is defined for the Euclidean plane but one can also consider the system in the hyperbolic plane or in other spaces that suitably generalize the plane. Outer billiards differs from a usual dynamical billiard in that it deals with a discrete sequence of moves outside the shape rather than inside of it.

The outer Billiards Map

Let P be a convex shape in the plane. Given a point x0 outside P, there is typically a unique point x1 (also outside P) so that the line segment connecting x0 to x1 is tangent to P at its midpoint and a person walking from x0 to x1 would see P on the right. (See Figure.) The map F: x0 -> x1 is called the outer billiards map.

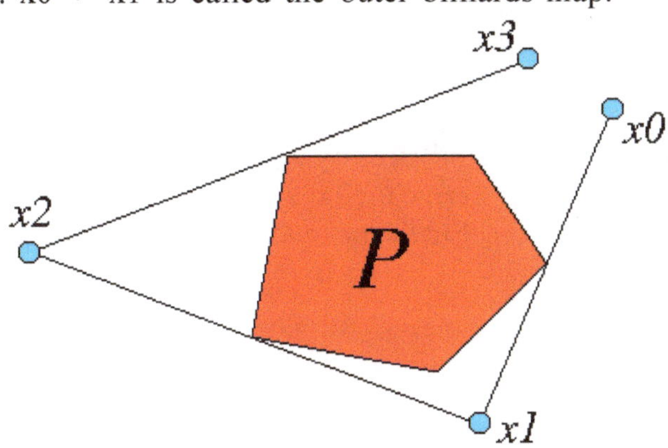

Outer billiards defined relative to a pentagon

The inverse (or backwards) outer billiards map is also defined, as the map x1 -> x0. One gets the inverse map simply by replacing the word right by the word left in the definition given above. The figure shows the situation in the Euclidean plane, but the definition in the hyperbolic plane is essentially the same.

Orbits

An outer billiards orbit is the set of all iterations of the point, namely ... x0 <--> x1 <--> x2 <--> x3 ... That is, start at x0 and iteratively apply both the outer billiards map and the backwards outer billiards map. When P is a strictly convex shape, such as an ellipse, every point in the exterior of P has a well defined orbit. When P is a polygon, some points might not have well-defined orbits, on account of the potential ambiguity of choosing the midpoint of the relevant tangent line. Nevertheless, in the polygonal case, almost every point has a well-defined orbit.

- An orbit is called periodic if it eventually repeats.
- An orbit is called aperiodic (or non-periodic) if it is not periodic.
- An orbit is called bounded (or stable) if some bounded region in the plane contains the whole orbit.
- An orbit is called unbounded (or unstable) if it is not bounded.

Higher-Dimensional Spaces

Defining an outer billiards system in a higher-dimensional space is beyond the scope of this article. Unlike the case of ordinary billiards, the definition is not straightforward. One natural setting for the map is a complex vector space. In this case, there is a natural choice of line tangent to a convex body at each point. One obtains these tangents by starting with the normals and using the complex structure to rotate 90 degrees. These distinguished tangent lines can be used to define the outer billiards map roughly as above. See S. Tabachnikov's book (cited in the references) for details. Most people attribute the introduction of outer billiards to Bernhard Neumann in the late 1950s, though it seems that a few people cite an earlier construction in 1945, due to M. Day. Jürgen Moser popularized the system in the 1970s as a toy model for celestial mechanics. This system has been studied classically in the Euclidean plane, and more recently in the hyperbolic plane. One can also consider higher-dimensional spaces, though no serious study has yet been made. Bernhard Neumann informally posed the question as to whether or not one can have unbounded orbits in an outer billiards system, and Moser put it in writing in 1973. Sometimes this basic question has been called the Moser-Neumann question. This question, originally posed for shapes in the Euclidean plane and solved only recently, has been a guiding problem in the field.

Moser-Neumann Question

Bounded orbits in the Euclidean plane

In the 70's, Jürgen Moser sketched a proof, based on K.A.M. theory, that outer billiards relative to a 6-times-differentiable shape of positive curvature has all orbits bounded. In 1982, Raphael Douady gave the full proof of this result. A big advance in the polygonal case came over a period of several years when three teams of authors, Vivaldi-Shaidenko (1987), Kolodziej (1989), and Gutkin-Simanyi (1991), each using different methods, showed that outer billiards relative to a quasirational polygon has all orbits bounded. The notion of quasirational is technical (see references) but it includes the class of regular polygons and convex rational polygons, namely those convex polygons whose vertices have rational coordinates. In the case of rational polygons, all the orbits are periodic. In 1995, Tabachnikov showed that outer billiards for the regular pentagon has some aperiodic orbits, thus clarifying the distinction between the dynamics in the rational and regular cases. In 1996, Boyland showed that outer billiards relative to some shapes can have orbits which accumulate on the shape. In 2005, D. Genin showed that all orbits are bounded when the shape is a trapezoid, thus showing that quasirationality is not a necessary condition for the system to have all orbits bounded. (Not all trapezoids are quasirational.)

Unbounded Orbits in the Euclidean Plane

In 2007, R. E. Schwartz showed that outer billiards has some unbounded orbits when defined relative to the Penrose Kite, thus answering the original Moser-Neumann question in the affirmative. The Penrose kite is the convex quadrilateral from the kites-and-darts Penrose tilings. Subsequently, Schwartz showed that outer billiards has unbounded orbits when defined relative to any irrational kite. An irrational kite is a quadrilateral with the following property: One of the diagonals of the quadrilateral divides the region into two triangles of equal area and the other diagonal divides the region into two triangles whose areas are not rational multiples of each other. In 2008, Dolgopyat-Fayad showed that outer billiards defined relative to the semidisk has unbounded orbits. The semidisk is the region one gets by cutting a disk in half. The proof of Dolgopyat-Fayad is robust, and also works for regions obtained by cutting a disk nearly in half, when the word nearly is suitably interpreted.

Unbounded Orbits in the Hyperbolic Plane

In 2003, Dogru and Tabachnikov showed that all orbits are unbounded for a certain class of convex polygons in the hyperbolic plane. The authors call such polygons large. (See the reference for the definition.) Dogru and Otten then extended this work in 2011 by specifying the conditions under which a regular polygonal table in the hyperbolic plane have all orbits unbounded, that is, are large.

Existence of Periodic Orbits

In ordinary polygonal billiards, the existence of periodic orbits is a major unsolved problem. For instance, it is unknown if every triangular shaped table has a periodic billiard path. More progress has been made for outer billiards, though the situation is far from well-understood. As mentioned above, all the orbits are periodic when the system is defined relative to a convex rational polygon in the Euclidean plane. Moreover, it is a recent theorem of C. Culter (written up by S. Tabachnikov) that outer billiards relative to any convex polygon has periodic orbits—in fact a periodic orbit outside of any given bounded region.

REFERENCES

1. Vladimir I. Arnold; A. Avez (1967). Problèmes Ergodiques de la Mécanique Classique (in French). Paris: Gauthier-Villars.; English translation: V. I. Arnold; A. Avez (1968). Ergodic Problems in Classical Mechanics. New York: Benjamin.
2. Franks, John M (October 1977). "Invariant sets of hyperbolic toral automorphisms". American Journal of Mathematics. The Johns Hopkins University Press. 99 (5): 1089–1095. doi:10.2307/2374001. ISSN 0002-9327.
3. «Sloane's A004146". The On-Line Encyclopedia of Integer Sequences. OEIS Foundation.
4. Dyson, Freeman John; Falk, Harold (1992). "Period of a Discrete Cat Mapping". The American Mathematical Monthly. Mathematical Association of America. 99 (7): 603–614. doi:10.2307/2324989. ISSN 0002-9890. JSTOR 2324989.
5. Blanchard, P.; Devaney, R. L.; Hall, G. R. (2006). Differential Equations. London: Thompson. pp. 96–111. ISBN 0-495-01265-3.
6. Henri Poincaré. "L'Équilibre d'une masse fluide animée d'un mouvement de rotation". Acta Mathematica, vol.7, pp. 259-380, Sept 1885.
7. Gao, J.; Delos, J. B. (1997). "Quantum manifestations of bifurcations of closed orbits in the photoabsorption spectra of atoms in electric fields". Phys. Rev. A. 56 (1): 356–364. Bibcode:1997PhRvA..56..356G. doi:10.1103/PhysRevA.56.356.
8. Peters, A. D.; Jaffé, C.; Delos, J. B. (1994). "Quantum Manifestations of Bifurcations of Classical Orbits: An Exactly Solvable Model". Phys. Rev. Lett. 73 (21): 2825–2828. Bibcode:1994PhRvL..73.2825P. doi:10.1103/PhysRevLett.73.2825. PMID 10057205.
9. Courtney, Michael; Jiao, Hong; Spellmeyer, Neal; Kleppner, Daniel; Gao, J.; Delos, J. B.; et al. (1995). "Closed Orbit Bifurcations in Continuum Stark Spectra". Phys. Rev. Lett. 74 (9): 1538–1541. Bibcode:1995PhRvL..74.1538C. doi:10.1103/PhysRevLett.74.1538. PMID 10059054.
10. Founargiotakis, M.; Farantos, S. C.; Skokos, Ch.; Contopoulos, G. (1997). "Bifurcation diagrams of periodic orbits for unbound molecular systems: FH2". Chemical Physics Letters. 277 (5–6): 456–

464. Bibcode:1997CPL...277..456F. doi:10.1016/S0009-2614(97)00931-7.

11. Monteiro, T. S. & Saraga, D. S. (2001). "Quantum Wells in Tilted Fields:Semiclassical Amplitudes and Phase Coherence Times". Foundations of Physics. 31 (2): 355–370. doi:10.1023/A:1017546721313.

12. Wieczorek, S.; Krauskopf, B.; Simpson, T. B. & Lenstra, D. (2005). "The dynamical complexity of optically injected semiconductor lasers". Physics Reports. 416 (1–2): 1–128. Bibcode:2005PhR...416....1W. doi:10.1016/j.physrep.2005.06.003.

13. Stamatiou, G. & Ghikas, D. P. K. (2007). "Quantum entanglement dependence on bifurcations and scars in non-autonomous systems. The case of quantum kicked top". Physics Letters A. 368 (3–4): 206–214. arXiv:quant-ph/0702172. Bibcode:2007PhLA..368..206S. doi:10.1016/j.physleta.2007.04.003.

14. Galan, J.; Freire, E. (1999). "Chaos in a Mean Field Model of Coupled Quantum Wells; Bifurcations of Periodic Orbits in a Symmetric Hamiltonian System". Reports on Mathematical Physics. 44 (1–2): 87–94. Bibcode:1999RpMP...44...87G. doi:10.1016/S0034-4877(99)80148-7.

15. Kleppner, D.; Delos, J. B. (2001). "Beyond quantum mechanics: Insights from the work of Martin Gutzwiller". Foundations of Physics. 31 (4): 593–612. doi:10.1023/A:1017512925106.

16. Gutzwiller, Martin C. (1990). Chaos in Classical and Quantum Mechanics. New York: Springer-Verlag. ISBN 0-387-97173-4.

CHAPTER 3

Controller

Chapter Contents

- **Introduction**
- **Input And Control Variables**
- **Types Of Controlling System**
- **Types Of Controller**

Sometimes, the control element has only two position either it is fully closed or fully open. This control element does not operate at any intermediate position, i.e. partly open or partly closed position. The control system made for controlling such elements, is known as on off control theory. In this control system, when process variable changes and crosses certain preset level, the output valve of the system is suddenly fully opened and gives 100 % output.Generally in on off control system, the output causes change in process variable. Hence due to effect of output, the process variable again starts changing but in reverse direction. During this change, when process variable crosses certain predetermined level, the output valve of the system is immediately closed and output is suddenly reduced to 0 %. As there is no output, the process variable again starts changing in its normal direction. When it crosses the preset level, the output valve of the system is again fully open to give 100 % output. This cycle of closing and opening of output valve continues till the said on-off control system is in operation. A very common example of on-off control theory is fan controlling scheme of transformer cooling system.

INTRODUCTION

In control theory, a **controller** is a device, historically using mechanical, hydraulic, pneumatic or electronic techniques often in combination, but more recently in the form of a microprocessor or computer, which monitors and physically alters the operating conditions of a given dynamical system.[1] [2] Typical applications of controllers are to hold settings for temperature, pressure, flow or speed.

INPUT AND CONTROL VARIABLES

A system can either be described as a MIMO system, having multiple inputs and outputs, therefore requiring more than one controller; or a SISO system, consisting of a single input and single output, hence having only a single controller. Depending on the set-up of the physical (or non-physical) system, adjusting the system's input variable (assuming it is SISO) will affect the operating parameter, otherwise known as the controlled output variable. Upon receiving the error signal that marks the disparity between the desired value (setpoint) and the actual output value, the controller will then attempt to regulate controlled output behaviour. The controller achieves this by either attenuating or amplifying the input signal to the plant so that the output is returned to the setpoint. For example, a simple feedback control system, such as the one shown on the right, will generate an error signal that›s mathematically depicted as the difference between the setpoint value and the output value, r-y.

This signal describes the magnitude by which the output value deviates from the setpoint. The signal is subsequently sent to the controller C which then interprets and adjusts for the discrepancy. If the plant is a physical one, the inputs to the system are regulated by means of actuators.

A thermostat on a heater is an example of open loop control that is on or off. A temperature sensor turns the heat source on if the temperature falls below the set point and turns the heat source off when the set point is reached. There is no measurement of the difference between the set point and the measured temperature (e.g. no error measurement) and no adjustment to the rate at which heat is added other than all or none.

A familiar example of feedback control is cruise control on an automobile. Here speed is the **measured variable**. The operator (driver) adjusts the desired speed **set point** (e.g. 100 km/hr) and the controller monitors the speed sensor and compares the measured speed to the set point. Any deviations, such as changes in grade, drag, wind speed or even using a different grade of fuel (for example an ethanol blend) are corrected by the controller making a compensating adjustment to the fuel valve open position, which is the **manipulated variable**. The controller makes adjustments having information only about the error (magnitude, rate of change or cumulative error) although adjustments known as *tuning* are used to achieve stable control. The operation of such controllers is the subject of control theory.

System	Controlled Outputs Include	Controller	Desired Performance Includes
Aircraft	Course, pitch, roll, yaw	Autopilot	Maintain flight path on a safe and smooth trajectory
Furnace	Temperature	Temperature controller	Follow warm-up temperature profile, then maintain temperature

Water treatment	pH value of effluent	pH controller	Neutralize effluent to specified accuracy
Automobile	Speed	Cruise control	Attain, then maintain selected speed without undue fuel consumption

The notion of controllers can be extended to more complex systems. In the natural world, individual organisms also appear to be equipped with controllers that assure the homeostasis necessary for survival of each individual. Both human-made and natural systems exhibit collective behaviors amongst individuals in which the controllers seek some form of equilibrium

TYPES OF CONTROLLING SYSTEM

In control theory there are two basic types of control: feedback and feed-forward.

Feedback

The input to a feedback controller is the same as what it is trying to control - the controlled variable is "fed back" into the controller. The thermostat of a house is an example of a feedback controller. This controller relies on measuring the controlled variable, in this case the temperature of the house, and then adjusting the output, whether or not the heater is on. However, feedback control usually results in intermediate periods where the controlled variable is not at the desired set-point. With the thermostat example, if the door of the house were opened on a cold day, the house would cool down. After it fell below the desired temperature (set-point), the heater would kick on, but there would be a period when the house was colder than desired.

Feed-forward

Feed-forward control can avoid the slowness of feedback control. With feed-forward control, the disturbances are measured and accounted for before they have time to affect the system. In the house example, a feed-forward system may measure the fact that the door is opened and automatically turn on the heater before the house can get too cold. The difficulty with feed-forward control is that the effect of the disturbances on the system must be accurately predicted, and there must not be any unmeasured disturbances. For instance, if a window were opened that was not being measured, the feed-forward-controlled thermostat might still let the house cool down.

To achieve the benefits of feedback control (controlling unknown disturbances and not having to know exactly how a system will respond to disturbances) *and* the benefits of feed-forward control (responding to disturbances before they can affect the system), there are combinations of feedback and feed-forward that can be used.

Examples

Some examples of where feedback and feed-forward control can be used together are dead-time compensation, and inverse response compensation. **Dead-time compensation** is used to control devices that take a long time to show any change to a change in input, for example, change in composition of flow through a long pipe. A dead-time compensation control uses an element (also called a Smith predictor) to predict how changes made now by the controller will affect the controlled variable in the future. The controlled variable is also measured and used in feedback control. **Inverse response compensation** involves controlling systems where a change at first affects the measured variable one way but later affects it in the opposite way. An example would be eating candy. At first it will give you lots of energy, but later you will be very tired. As can be imagined, it is difficult to control this system with feedback alone, therefore a predictive feed-forward element is necessary to predict the reverse effect that a change will have in the future.

TYPES OF CONTROLLER

Most control valve systems in the past were implemented using mechanical systems or solid state electronics. Pneumatics were often utilized to transmit information and control using pressure. However, most modern industrial control systems now rely on computers for the **industrial controller**. Obviously it is much easier to implement complex control algorithms on a computer than using a mechanical system.

For feedback controllers there are a few simple types. The most simple is like the thermostat that just turns the heat on if the temperature falls below a certain value and off it exceeds a certain value (on-off control).

Another simple type of controller is a proportional controller. With this type of controller, the controller output (control action) is proportional to the error in the measured variable.

In feedback control, it is standard to define the error as the difference between the desired value (setpoint) y_s and the current value (measured) y. If the error is large, then the control action is large. Mathematically:

$$u(t) = K_c * e(t) + u_0$$

where

$u(t)$ represents the control action (controller output),

$e(t) = y_s(t) - y(t)$ represents the error,

K_c represents the controller's gain, and

U_0 represents the steady state control action (bias) necessary to maintain the variable at the steady state when there is no error.

It is important that the control action u counteracts the change in the controlled variable-v. There are then two cases depending on the sign of the process gain.

In the first case the process gain is positive, so an increase in the controlled variable (measurement) y requires a decrease in the control action u (reverse-acting control). In this case the controller gain K_c is positive, because the standard definition of the error already contains a negative sign for y.

In the second case the process gain is negative, so an increase in the controlled variable (measurement) y requires an increase in the control action u (direct-acting control). In this case the controller gain K_c is negative.

A typical example of a reverse-acting system is control of temperature y by use of steam u. In this case the process gain is positive, so if the temperature increases, the steam flow must be decreased to maintain the desired temperature. Conversely, a typical example of a direct-acting system is control of temperature using cooling water. In this case the process gain is negative, so if the temperature increases, the cooling water flow must be increased to maintain the desired temperature.

Although proportional control is simple to understand, it has drawbacks. The largest problem is that for most systems it will never entirely remove error. This is because when error is 0 the controller only provides the steady state control action so the system will settle back to the original steady state (which is probably not the new set point that we want the system to be at). To get the system to operate near the new steady state, the controller gain, Kc, must be very large so the controller will produce the required output when only a very small error is present. Having large gains can lead to system instability or can require physical impossibilities like infinitely large valves.

Alternates to proportional control are proportional-integral (PI) control and proportional-integral-derivative (PID) control. PID control is commonly used to implement closed-loop control.

Open-loop control can be used in systems sufficiently well-characterized as to predict what outputs will necessarily achieve the desired states. For example, the rotational velocity of an electric motor may be well enough characterized for the supplied voltage to make feedback unnecessary.

The drawback of open-loop control is that it requires perfect knowledge of the system (i.e. one knows exactly what inputs to give in order to get the desired output), and it assumes there are no disturbances to the system.

REFERENCES

1. ACODUASIS IPS-2001-42068 : ACODUASIS Project web-site
2. CORDIS Issue n. 44 – September 2003, *Technology opportunities today*, page 16: "*EICASLAB: A family of CAE tools using automated algorithm generation to design control systems*". Published by European Commission – Innovation, on line
3. EVCA Barometer April 2006, page 5: "*An easy to use tool for automated control systems*", on line
4. CORDIS - ICT results: Results that lead the way: "*An easy-to-use tool for automated control systems*", Published by European Commission, on line
5. Prof. Francesco Donati (Politecnico of Torino, Italy): *"The innovative methodology and the ACODUASIS Project"*, ACODUASIS Workshop "One Step further in Automatic Control Design", Torino (Italy), 3 October 2005, on line

6. Gabriella Caporaletti (EICAS Automazione, Italy): *"The ACODUASIS Project: A professional software tool supporting the control design in robotics"*, 6th International Conference on Climbing and Walking Robots And the Support Technologies for Mobile Machines. CLAWAR 2003 September 17–19, 2003, Catania, Italy
7. ARFLEX Project IST-NMP2-016880 : ARFLEX Project web-site/
8. PISA Project NMP2-CT-2006-026697 PISA Project web-site
9. HI-CEPS Project TIP5-CT-2006-031373: HI-CEPS Project web-site
10. ERSEC Project FP7 247955: ERSEC Project web-site:
11. Gabriella Caporaletti (EICAS Automazione, Italy), Rui Neves da Silva and Maria Marques (UNINOVA, Portugal): «*Advanced Automated Algorithm Generation Software in the Control of Solar Plant*" - MIC 2004 Twenty-Third IASTED International Conference on Modelling, Identification and Control, abstract on line
12. Kerscher, Zoellner and Dillman (University of Karlsruhe, Germany), Stella and Caporaletti (EICAS Automazione, Italy):»*Model and Control of joints driven by fluidic muscles with the help of advanced automatic algorithm generation software*"- CLAWAR 2005 8th International Conference on Climbing and Walking Robots and the Support Technologies for Mobile Machines
13. Kay Ch. Fuerstenberg (IBEO Automobile Sensor GmbH, Germany), Pierre Baraud (Peugeot Citroën Automobile, France), Gabriella Caporaletti (EICAS Automazione,Italy), Silvia Citelli (Fiat Research Center,Italy), Zafrir Eitan (TAMAM/IAI, Israel), Ulrich Lages (IBEO Automobile Sensor GmbH, Germany), Christophe Lavergne (Renault SA, France) *Development of a Pre-crash sensorial system: The CHAMELEON Project*, on line
14. A. Bottero and D. Martinello (COMAU Robotics, Italy):*Industrial robot simulation models for control design and analysis purposes*, ACODUASIS Workshop "One Step further in Automatic Control Design", Torino (Italy), 3 October 2005, on line
15. F. Motto and A. Ramoino (EICAS Automazione, Italy), A. Bottero and D. Martinello (COMAU Robotics,Italy: *Industrial robots control with EICASLAB approach: industrial prototyping and experimentation results*, ACODUASIS Workshop "One Step further in Automatic Control Design", Torino (Italy), 3 October 2005, on line
16. J. Fottner (MIAS, Germany), T. Kerscher (University of Karlsruhe, Germany), G. di Gropello and A. Stella (EICAS Automazione, Italy): *Modelling and Control of Automated Guided Vehicles (AGVs) for the transport of meals, laundry and waste in the healthcare domain*, ACODUASIS Workshop "One Step further in Automatic Control Design", Torino (Italy), 3 October 2005, on line
17. G. Caporaletti and A. Stella (EICAS Automazione, Italy), P. Pina (UNINOVA,Portugal), V. Abadie (CYBERNETIX, France): *"Control of a hydraulic servoactuator using an automated algorithm generator"*, ACODUASIS Workshop "One Step further in Automatic Control Design", Torino (Italy), 3 October 2005, on line
18. Y. Dodeman and N. Moisan (IPSIS, France), G. di Gropello (EICAS Automazione, Italy):*"Synthesis of multivariable control of a thermic power plant"*, ACODUASIS Workshop "One Step further in Automatic Control Design", Torino (Italy), 3 October 2005, on line
19. Prof. R. Bucher and K.Kaufmann – (SUPSI, Switzerland):»*Rapid Control Prototyping with EICASLAB and Linux RTAI*", ACODUASIS Workshop "One Step further in Automatic Control Design", Torino (Italy), 3 October 2005, on line

20. Prof. Silvano Balemi (University of Applied Sciences of Southern Switzerland, Lugano-Manno, Switzerland): «*Rapid Controller Prototyping Platform for Precision Applications*", Proceedings of the 6th euspen International Conference – Baden bei Wien - May 2006, on line

21. Paola Donati and Francesco Donati:" Modelling and Forecasting the Yield Curve under Model uncertainty, by Paola Donati and Francesco Donati", Working Paper Series 917, European Central Bank (ECB), on line

22. Paola Donati:" Monetary Policy Effectiveness in Times of Crisis: Evidence from the Euro Area Money Market", European Central Bank (ECB) workshop December 2009, on line

23. Prof. F. Donati , Prof. D. Carlucci: «*Control of norm of uncertain systems*", IEEE Transactions on Automatic Control, vol.20-AC, 1975, pp.792- 795

24. Prof. F. Donati , Prof. M. Vallauri: «*Guaranteed control of almost-linear plants*", IEEE Transactions on Automatic Control, vol. 29- AC, 1984, pp. 34-41

CHAPTER 4

Process Control

Chapter Contents

Control in process industries refers to the regulation of all aspects of the process. Precise control of level, temperature, pressure and flow is important in many process applications. This module introduces you to control in process industries, explains why control is important, and identifies different ways in which precise control is ensured.

As you proceed through the module, answer the questions in the activities column on the right side of each page. Also, note the application boxes (double-bordered boxes) located throughout the module. Application boxes provide key information about how you may use your baseline knowledge in the field. When you see the workbook exercise graphic at the bottom of a page, go to the workbook to complete the designated exercise before moving on in the module. Workbook exercises help you measure your progress toward meeting each section's learning objectives.

INTRODUCTION

Process control is an engineering discipline that deals with architectures, mechanisms and algorithms for maintaining the output of a specific process within a desired range. For instance, the temperature of a chemical reactor may be controlled to maintain a consistent product output.

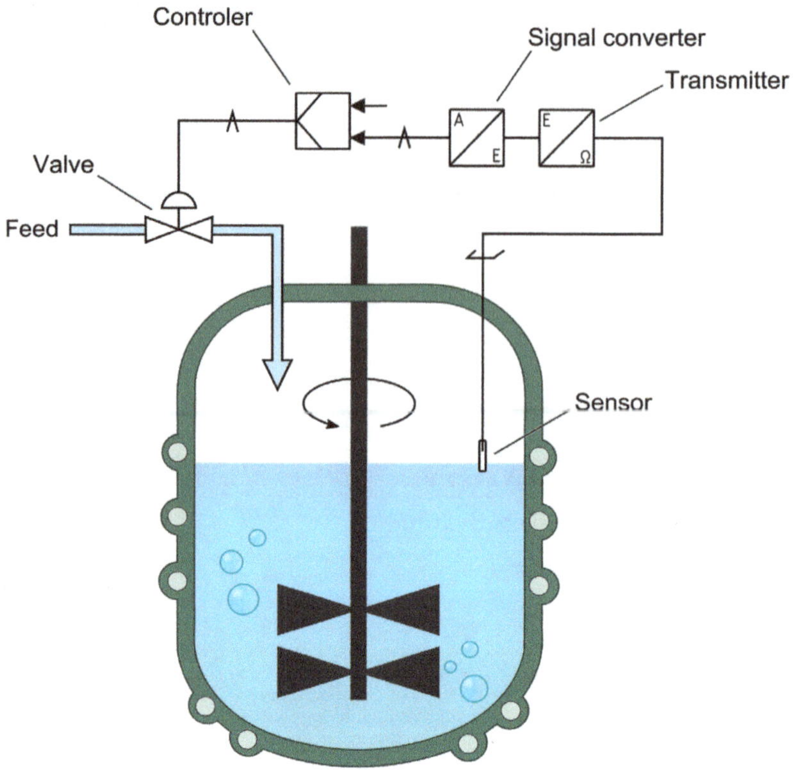

Example of control system of a continuous stirred-tank reactor.

Control panel of a nuclear reactor.

Process control is extensively used in industry and enables mass production of consistent products from continuously operated processes such as oil refining, paper manufacturing, chemicals, power plants and many others. Process control enables automation, by which a small staff of operating personnel can operate a complex process from a central control room.

BACKGROUND

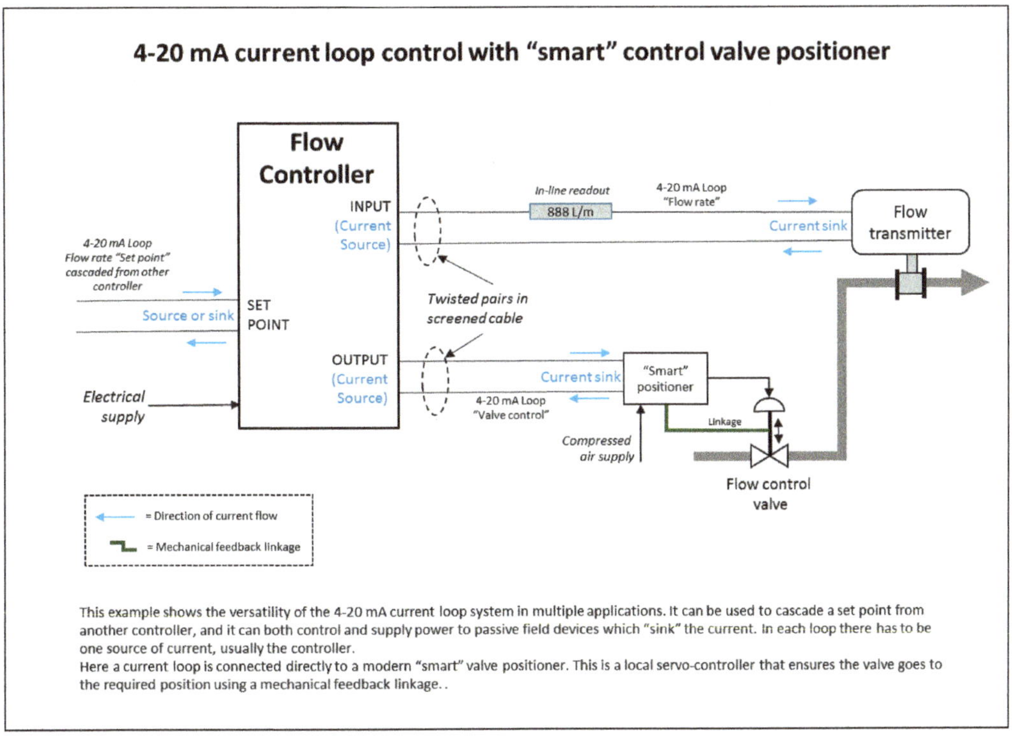

Example of a continuous flow control loop. Signalling is by industry standard 4-20 mA current loops, and a "smart" valve positioner ensures the control valve operates correctly.

Process control may either use feedback or it may be open loop. Control may also be continuous (automobile cruise control) or cause a sequence of discrete events, such as a timer on a lawn sprinkler (on/off) or controls on an elevator (logical sequence).

A thermostat on a heater is an example of control that is on or off. A temperature sensor turns the heat source on if the temperature falls below the set point and turns the heat source off when the set point is reached. There is no measurement of the difference between the set point and the measured temperature (e.g. no error measurement) and no adjustment to the rate at which heat is added other than all or none. A familiar example of feedback control is cruise control on an automobile. Here speed is the **measured**

variable. The operator (driver) adjusts the desired speed **set point** (e.g. 100 km/hr) and the controller monitors the speed sensor and compares the measured speed to the set point. Any deviations, such as changes in grade, drag, wind speed or even using a different grade of fuel (for example an ethanol blend) are corrected by the controller making a compensating adjustment to the fuel valve open position, which is the **manipulated variable**. The controller makes adjustments having information only about the error (magnitude, rate of change or cumulative error) although settings known as *tuning* are used to achieve stable control. The operation of such controllers is the subject of control theory.

A commonly used control device called a programmable logic controller, or a PLC, is used to read a set of digital and analog inputs, apply a set of logic statements, and generate a set of analog and digital outputs.

For example, if an adjustable valve were used to hold level in a tank the logical statements would compare the equivalent pressure at depth setpoint to the pressure reading of a sensor below the normal low liquid level and determine whether more or less valve opening was necessary to keep the level constant. A PLC output would then calculate an incremental amount of change in the valve position. Larger more complex systems can be controlled by a Distributed Control System (DCS) or SCADA system.

HIERARCHY OF PROCESS CONTROL

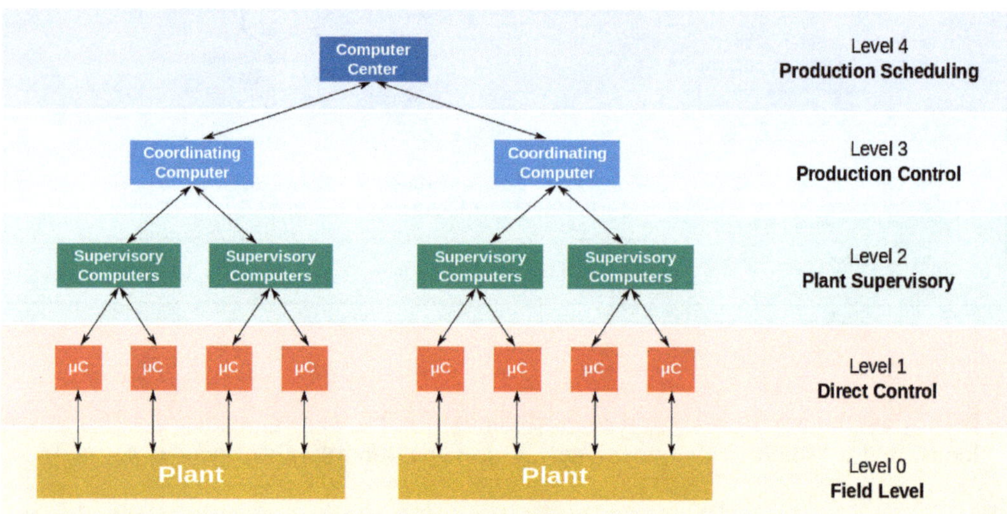

Functional levels of a manufacturing control operation.

The accompanying diagram is a general model which shows functional manufacturing levels in a large process using computerised control.

Referring to the diagram;

- Level 0 contains the field devices such as flow and temperature sensors, and final control elements, such as control valves

- Level 1 contains the industrialised Input/Output (I/O) modules, and their associated distributed electronic processors.
- Level 2 contains the supervisory computers, which collate information from processor nodes on the system, and provide the operator control screens.
- Level 3 is the production control level, which does not directly control the process, but is concerned with monitoring production and monitoring targets
- Level 4 is the production scheduling level.

TYPES OF PROCESSES USING PROCESS CONTROL

Processes can be characterized as one or more of the following forms:

- Discrete – Found in many manufacturing, motion and packaging applications. Robotic assembly, such as that found in automotive production, can be characterized as discrete process control. Most discrete manufacturing involves the production of discrete pieces of product, such as metal stamping.
- Batch – Some applications require that specific quantities of raw materials be combined in specific ways for particular durations to produce an intermediate or end result. One example is the production of adhesives and glues, which normally require the mixing of raw materials in a heated vessel for a period of time to form a quantity of end product. Other important examples are the production of food, beverages and medicine. Batch processes are generally used to produce a relatively low to intermediate quantity of product per year (a few pounds to millions of pounds).
- Continuous – Often, a physical system is represented through variables that are smooth and uninterrupted in time. The control of the water temperature in a heating jacket, for example, is an example of continuous process control. Some important continuous processes are the production of fuels, chemicals and plastics. Continuous processes in manufacturing are used to produce very large quantities of product per year (millions to billions of pounds).

Applications having elements of discrete, batch and continuous process control are often called *hybrid* applications.

EXAMPLES

- An anti-lock braking system (ABS) is a complex example, consisting of multiple inputs, conditions and outputs.
- Aircraft stability control is a highly complex example using multiple inputs and outputs.

ANTI-LOCK BRAKING SYSTEM

Symbol for ABS

ABS brakes on a BMW motorcycle

An anti-lock braking system or anti-skid braking system (ABS) is an automobile safety system that allows the wheels on a motor vehicle to maintain tractive contact with the road surface according to driver inputs while braking, preventing the wheels from locking up (ceasing rotation) and avoiding uncontrolled skidding. It is an automated system that uses the principles of threshold braking and cadence braking which were practiced by skillful drivers with previous generation braking systems. It does this at a much faster rate and with better control than many drivers could manage.

ABS generally offers improved vehicle control and decreases stopping distances on dry and slippery surfaces; however, on loose gravel or snow-covered surfaces, ABS can significantly increase braking distance, although still improving vehicle steering control.

Since initial widespread use in production cars, anti-lock braking systems have been improved considerably. Recent versions not only prevent wheel lock under braking, but also electronically control the front-to-rear brake bias. This function, depending on its specific capabilities and implementation, is known as electronic brakeforce distribution (EBD), traction control system, emergency brake assist, or electronic stability control (ESC).

Early Systems

The concept for ABS predates the modern systems that were introduced in the 1950s. In 1908, for example, J.E. Francis introduced his 'Slip Prevention Regulator for Rail Vehicles'. In 1920 the French automobile and aircraft pioneer Gabriel Voisin experimented with systems that modulated the hydraulic braking pressure on his aircraft brakes to reduce the risk of tyre slippage. These systems use a flywheel and valve attached to a hydraulic line that feeds the brake cylinders. The flywheel is attached to a drum that runs at the same speed as the wheel. In normal braking, the drum and flywheel should spin at the same speed. However, when a wheel slows down, then the drum would do the same, leaving the flywheel spinning at a faster rate. This causes the valve to open, allowing a small amount of brake fluid to bypass the master cylinder into a local reservoir, lowering the pressure on the cylinder and releasing the brakes. The use of the drum and flywheel meant the valve only opened when the wheel was turning. In testing, a 30% improvement in braking performance was noted, because the pilots immediately applied full brakes instead of slowly increasing pressure in order to find the skid point. An additional benefit was the elimination of burned or burst tires.

The first patented system was created by German engineer Karl Wessel in 1928. Wessel, however, never developed a working product and neither did Robert Bosch who produced a similar patent eight years later.

By the early 1950s, the Dunlop Maxaret anti-skid system was in widespread aviation use in the UK, with aircraft such as the Avro Vulcan and Handley Page Victor, Vickers Viscount, Vickers Valiant, English Electric Lightning, de Havilland Comet 2c, de Havilland Sea Vixen, and later aircraft, such as the Vickers VC10, Hawker Siddeley Trident, Hawker Siddeley 125, Hawker Siddeley HS 748 and derived British Aerospace ATP, and BAC One-Eleven being fitted with Maxaret as standard. Maxaret, while reducing braking distances by up to 30% in icy or wet conditions, also increased tyre life, and had the additional advantage of allowing take-offs and landings in conditions that would preclude flying at all in non-Maxaret equipped aircraft.

In 1958, a Royal Enfield Super Meteor motorcycle was used by the Road Research Laboratory to test the Maxaret anti-lock brake. The experiments demonstrated that anti-lock brakes can be of great value to motorcycles, for which skidding is involved in a high proportion of accidents. Stopping distances were reduced in most of the tests compared with locked wheel braking, particularly on slippery surfaces, in which the improvement could be as much as 30 percent. Enfield's technical director at the time, Tony Wilson-Jones, saw little future in the system, however, and it was not put into production by the company.

A fully mechanical system saw limited automobile use in the 1960s in the Ferguson P99 racing car, the Jensen FF, and the experimental all wheel drive Ford Zodiac, but saw no further use; the system proved expensive and unreliable.

The first fully electronic anti lock system was developed in the late 60s for the Concorde aircraft.

Modern Systems

Chrysler, together with the Bendix Corporation, introduced a computerized, three-channel, four-sensor all-wheel ABS called "Sure Brake" for its 1971 Imperial. It was available for several years thereafter, functioned as intended, and proved reliable. In 1970, Ford added an antilock braking system called "Sure-track" to the rear wheels of Lincoln Continentals as an option;it became standard in 1971. In 1971, General Motors introduced the "Trackmaster" rear-wheel only ABS as an option on their rear-wheel drive Cadillac models and the Oldsmobile Toronado. In the same year, Nissan offered an EAL (Electro Anti-lock System) developed by Japanese company Denso as an option on the Nissan President, which became Japan's first electronic ABS.

1971: Electronically controlled anti-skid brakes on Toyota Crown In 1972, four wheel drive Triumph 2500 Estates were fitted with Mullard electronic systems as standard. Such cars were very rare however and very few survive today.

Mercedes-Benz W116 became the first production car to use an electronic four-wheel multi-channel anti-lock braking system (ABS) from Bosch as an option from 1978 on.

In 1985 the Ford Scorpio was introduced to European market with a Teves electronic system throughout the range as standard. For this the model was awarded the coveted European Car of the Year Award in 1986, with very favourable praise from motoring journalists. After this success Ford began research into Anti-Lock systems for the rest of their range, which encouraged other manufacturers to follow suit.

In 1988, BMW introduced the first motorcycle with an electronic-hydraulic ABS: the BMW K100. Honda followed suit in 1992 with the launch of its first motorcycle ABS on the ST1100 Pan European. In 2007, Suzuki launched its GSF1200SA (Bandit) with an ABS. In 2005, Harley-Davidson began offering an ABS option on police bikes.

In 1993, Lincoln was the first automobile company to provide standard four-wheel anti-lock brakes on all of their vehicles.

Operation

The anti-lock brake controller is also known as the CAB (Controller Anti-lock Brake).

Typically ABS includes a central electronic control unit (ECU), four wheel speed sensors, and at least two hydraulic valves within the brake hydraulics. The ECU constantly monitors the rotational speed of each wheel; if it detects a wheel rotating significantly slower than the others, a condition indicative of impending wheel lock, it actuates the valves to reduce hydraulic pressure to the brake at the affected wheel, thus reducing the braking force on that wheel; the wheel then turns faster. Conversely, if the ECU detects a wheel turning significantly faster than the others, brake hydraulic pressure to the wheel is increased so the braking force is reapplied, slowing down the wheel. This process is repeated continuously and can be detected by the driver via brake pedal pulsation. Some anti-lock systems can apply or release braking pressure 15 times per second. Because of this, the wheels of cars equipped with ABS are practically impossible to lock even during panic braking in extreme conditions.

The ECU is programmed to disregard differences in wheel rotative speed below a critical threshold, because when the car is turning, the two wheels towards the center of the curve turn slower than the outer two. For this same reason, a differential is used in virtually all roadgoing vehicles.

If a fault develops in any part of the ABS, a warning light will usually be illuminated on the vehicle instrument panel, and the ABS will be disabled until the fault is rectified.

Modern ABS applies individual brake pressure to all four wheels through a control system of hub-mounted sensors and a dedicated micro-controller. ABS is offered or comes standard on most road vehicles produced today and is the foundation for electronic stability control systems, which are rapidly increasing in popularity due to the vast reduction in price of vehicle electronics over the years.

Modern electronic stability control systems are an evolution of the ABS concept. Here, a minimum of two additional sensors are added to help the system work: these are a steering wheel angle sensor, and a gyroscopic sensor. The theory of operation is simple: when the gyroscopic sensor detects that the direction taken by the car does not coincide with what the steering wheel sensor reports, the ESC software will brake the necessary individual wheel(s) (up to three with the most sophisticated systems), so that the vehicle goes the way the driver intends. The steering wheel sensor also helps in the operation of Cornering Brake Control (CBC), since this will tell the ABS that wheels on the inside of the curve should brake more than wheels on the outside, and by how much.

ABS equipment may also be used to implement a traction control system (TCS) on acceleration of the vehicle. If, when accelerating, the tire loses traction, the ABS controller can detect the situation and take suitable action so that traction is regained. More sophisticated versions of this can also control throttle levels and brakes simultaneously.

The speed sensors of ABS are sometimes used in indirect tire pressure monitoring system (TPMS), which can detect under-inflation of tire(s) by difference in rotational speed of wheels.

Components

There are four main components of ABS: wheel speed sensors, valves, a pump, and a controller.

Speed sensors

A speed sensor is used to determine the acceleration or deceleration of the wheel. These sensors use a magnet and a Hall effect sensor, or a toothed wheel and an electromagnetic coil to generate a signal. The rotation of the wheel or differential induces a magnetic field around the sensor. The fluctuations of this magnetic field generate a voltage in the sensor. Since the voltage induced in the sensor is a result of the rotating wheel, this sensor can become inaccurate at slow speeds. The slower rotation of the wheel can cause inaccurate fluctuations in the magnetic field and thus cause inaccurate readings to the controller.

Valves

There is a valve in the brake line of each brake controlled by the ABS. On some systems, the valve has three positions:

- In position one, the valve is open; pressure from the master cylinder is passed right through to the brake.
- In position two, the valve blocks the line, isolating that brake from the master cylinder. This prevents the pressure from rising further should the driver push the brake pedal harder.

- In position three, the valve releases some of the pressure from the brake.

The majority of problems with the valve system occur due to clogged valves. When a valve is clogged it is unable to open, close, or change position. An inoperable valve will prevent the system from modulating the valves and controlling pressure supplied to the brakes.

Pump

The pump in the ABS is used to restore the pressure to the hydraulic brakes after the valves have released it. A signal from the controller will release the valve at the detection of wheel slip. After a valve release the pressure supplied from the user, the pump is used to restore a desired amount of pressure to the braking system. The controller will modulate the pumps status in order to provide the desired amount of pressure and reduce slipping.

Controller

The controller is an ECU type unit in the car which receives information from each individual wheel speed sensor, in turn if a wheel loses traction the signal is sent to the controller, the controller will then limit the brake force (EBD) and activate the ABS modulator which actuates the braking valves on and off.

USE

There are many different variations and control algorithms for use in ABS. One of the simpler systems works as follows:

- The controller monitors the speed sensors at all times. It is looking for decelerations in the wheel that are out of the ordinary. Right before a wheel locks up, it will experience a rapid deceleration. If left unchecked, the wheel would stop much more quickly than any car could. It might take a car five seconds to stop from 60 mph (96.6 km/h) under ideal conditions, but a wheel that locks up could stop spinning in less than a second.
- The ABS controller knows that such a rapid deceleration is impossible, so it reduces the pressure to that brake until it sees an acceleration, then it increases the pressure until it sees the deceleration again. It can do this very quickly, before the tire can actually significantly change speed. The result is that the tire slows down at the same rate as the car, with the brakes keeping the tires very near the point at which they will start to lock up. This gives the system maximum braking power.
- This replaces the need to manually pump the brakes while driving on a slippery or a low traction surface, allowing to steer even in most emergency braking conditions.
- When the ABS is in operation the driver will feel a pulsing in the brake pedal; this comes from the rapid opening and closing of the valves. This pulsing also tells the driver that the ABS has been triggered. Some ABS systems can cycle up to 16 times per second.

Brake types

Anti-lock braking systems use different schemes depending on the type of brakes in use. They can be differentiated by the number of channels: that is, how many valves that are individually controlled—and the number of speed sensors.

Four-channel, four-sensor ABS

There is a speed sensor on all four wheels and a separate valve for all four wheels. With this setup, the controller monitors each wheel individually to make sure it is achieving maximum braking force.

Three-channel, four-sensor ABS

There is a speed sensor on all four wheels and a separate valve for each of the front wheels, but only one valve for both of the rear wheels. Older vehicles with four-wheel ABS usually use this type.

Three-channel, three-sensor ABS

This scheme, commonly found on pickup trucks with four-wheel ABS, has a speed sensor and a valve for each of the front wheels, with one valve and one sensor for both rear wheels. The speed sensor for the rear wheels is located in the rear axle. This system provides individual control of the front wheels, so they can both achieve maximum braking force. The rear wheels, however, are monitored together; they both have to start to lock up before the ABS will activate on the rear. With this system, it is possible that one of the rear wheels will lock during a stop, reducing brake effectiveness. This system is easy to identify, as there are no individual speed sensors for the rear wheels.

Two-channel, four sensor ABS

This system, commonly found on passenger cars from the late ‘80s through the mid 1990s, uses a speed sensor at each wheel, with one control valve each for the front and rear wheels as a pair. If the speed sensor detects lock up at any individual wheel, the control module pulses the valve for both wheels on that end of the car.

One-channel, one-sensor ABS

This system is commonly found on pickup trucks, SUVs, and vans with rear-wheel ABS. It has one valve, which controls both rear wheels, and one speed sensor, located in the rear axle. This system operates the same as the rear end of a three-channel system. The rear wheels are monitored together and they both have to start to lock up before the ABS kicks in. In this system it is also possible that one of the rear wheels will lock, reducing brake effectiveness. This system is also easy to identify, as there are no individual speed sensors for any of the wheels.

Effectiveness

A 2004 Australian study by Monash University Accident Research Centre found that ABS:

- Reduced the risk of multiple vehicle crashes by 18 percent,
- Increased the risk of run-off-road crashes by 35 percent.

On high-traction surfaces such as bitumen, or concrete, many (though not all) ABS-equipped cars are able to attain braking distances better (i.e. shorter) than those that would be possible without the benefit of ABS. In real world conditions, even an alert and experienced driver without ABS would find it difficult to match or improve on the performance of a typical driver with a modern Anti-lock Braking System-equipped vehicle. ABS reduces chances of crashing, and/or the severity of impact. The recommended technique for non-expert drivers in an ABS-equipped car, in a typical full-braking emergency, is to press the brake pedal as firmly as possible and, where appropriate, to steer around obstructions. In such situations, ABS will significantly reduce the chances of a skid and subsequent loss of control.

In gravel, sand and deep snow, ABS tends to increase braking distances. On these surfaces, locked wheels dig in and stop the vehicle more quickly. ABS prevents this from occurring. Some ABS calibrations reduce this problem by slowing the cycling time, thus letting the wheels repeatedly briefly lock and unlock. Some vehicle manufacturers provide an "off-road" button to turn ABS function off. The primary benefit of ABS on such surfaces is to increase the ability of the driver to maintain control of the car rather than go into a skid, though loss of control remains more likely on soft surfaces such as gravel or on slippery surfaces such as snow or ice. On a very slippery surface such as sheet ice or gravel, it is possible to lock multiple wheels at once, and this can defeat ABS (which relies on comparing all four wheels, and detecting individual wheels skidding). Availability of ABS relieves most drivers from learning threshold braking.

A June 1999 National Highway Traffic Safety Administration (NHTSA) study found that ABS increased stopping distances on loose gravel by an average of 27.2 percent.

According to the NHTSA,

"ABS works with your regular braking system by automatically pumping them. In vehicles not equipped with ABS, the driver has to manually pump the brakes to prevent wheel lockup. In vehicles equipped with ABS, your foot should remain firmly planted on the brake pedal, while ABS pumps the brakes for you so you can concentrate on steering to safety."

When activated, some earlier ABS systems caused the brake pedal to pulse noticeably. As most drivers rarely or do not brake hard enough to cause brake lock-up, and drivers typically do not read the vehicle's owners manual, this may not be noticeable until an emergency. Some manufacturers have therefore implemented a brake assist system that determines that the driver is attempting a "panic stop" (by detecting that the brake pedal was depressed very fast, unlike a normal stop where the pedal pressure would usually be gradually increased, Some systems additionally monitor the rate at the accelerator was released) and the system automatically increases braking force where not enough pressure is applied. Hard or panic braking on bumpy surfaces, because of the bumps causing the speed of the wheel(s) to become erratic may also trigger the ABS, sometimes causing the system to enter its ice mode, where the system severely limits maximum available braking power. Nevertheless, ABS significantly improves safety and control for drivers in most on-road situations.

Anti-lock brakes are the subject of some experiments centred around risk compensation theory, which asserts that drivers adapt to the safety benefit of ABS by driving more aggressively. In a Munich study, half a fleet of taxicabs was equipped with anti-lock brakes, while the other half had conventional brake systems. The crash rate was substantially the same for both types of cab, and Wilde concludes this was due to drivers of ABS-equipped cabs taking more risks, assuming that ABS would take care of them, while the non-ABS drivers drove more carefully since ABS would not be there to help in case of a dangerous situation.

The Insurance Institute for Highway Safety released a study in 2010 that found motorcycles with ABS 37% less likely to be involved in a fatal crash than models without ABS.

Regulations

ABS are required on all new passenger cars sold in the EU since 2004. In the United States, the NHTSA has mandated ABS in conjunction with Electronic Stability Control under the provisions of FMVSS 126 as of September 1, 2013.

REFERENCES

1. "Toyota Prius c features safety anti-skid braking system (ABS)". Toyota Motor Corporation Australia. Retrieved 7 March 2016. Prius c is equipped with Toyota's Anti-skid Braking System (ABS).
2. Effectiveness of ABS and Vehicle Stability Control Systems" (PDF). Royal Automobile Club of Victoria. April 2004. Retrieved 2010-12-07.
3. Heißing, Bernd (2011), Chassis Handbook, Springer, ISBN 9783834897893, retrieved February 19, 2013
4. Cycle World, June 1992, ISSN 0011-4286, retrieved February 19, 2013
5. Jon Lawes (31 January 2014). Car Brakes: A Guide to Upgrading, Repair and Maintenance. Crowood. pp. 207–. ISBN 978-1-84797-675-8.
6. Flight International. 30 October 1953. pp. 587–588.
7. «Browse Flight's archive of Historic Aviation". Flightglobal.com. Retrieved 2014-08-26.
8. Reynolds, Jim (1990). Best of British Bikes. Patrick Stephens Ltd. ISBN 1-85260-033-0.
9. «Directory Index: Chrysler_and_Imperial/1972 Chrysler/1972_Imperial_Press_Kit". Oldcarbrochures.com. Retrieved 2014-08-26.
10. «Chrysler Imperial Sure Brake system description". ImperialClub.org.
11. «Directory Index: Lincoln/1970_Lincoln/1970 Lincoln Continental Brochure".

Oldcarbrochures.com. Retrieved 2014-08-26.

12. «Directory Index: Lincoln/1971_Lincoln/1971_Lincoln_Continental_Brochure". Oldcarbrochures.com. Retrieved 2014-08-26.
13. «Directory Index: Cadillac/1972_Cadillac/1972_Cadillac_Brochure_1". Oldcarbrochures.com. Retrieved 2014-08-26.
14. «History". We Love Cadillacs.
15. «1972, First Automotive Anti-lock Brake System (ABS)". gmheritagecenter.com.
16. «Directory Index: Oldsmobile/1971 Oldsmobile/album". Oldcarbrochures.com. Retrieved 2014-08-26.
17. «Electro antilock system (installed in Nissan President)". 240 Landmarks of Japanese Automotive Technology. Society of Automotive Engineers in Japan, Inc.
18. «TOYOTA MOTOR CORPORATION GLOBAL WEBSITE - 75 Years of TOYOTA - Technical Development - Chassis". toyota-global.com.
19. Anthony Slanda (15 July 2007). "1993 Lincoln safety ad"– via YouTube.
20. KI4CY (2003-02-13). "Ram Glossary of abbreviations and terms". Dodgeram.org. Archived from the original on 22 November 2010. Retrieved 2010-12-07.
21. Nice, Karim. How "Anti-Lock Brakes Work". howstuffworks. Retrieved October 2, 2010.
22. «ABS Frequently Asked Questions". ABS Education Alliance. 2004-05-03. Archived from the original on 23 May 2010. Retrieved 2009-10-22.
23. NHTSA Light Vehicle Antilock Brake System Research Program Task 4: A Test Track Study of Light Vehicle ABS Performance Over a Broad Range of Surfaces and Maneuvers, Jan 1999 PDF
24. Gerald J. S. Wilde (1994). "7. Remedy by engineering?". Psyc.queensu.ca. Retrieved 2010-12-07.
25. "Motorcycle ABS: Skepticism Debunked". Ultimate Motorcycling. 2012-05-16. Retrieved 2012-08-18.
26. Electronic Stability Control (ESC). nhtsa.gov.

CHAPTER 5

Applications of Control Theory

Chapter Contents

Control theory has its roots in the use of feedback as a means to regulate physical processes and mediate the effect of modeling uncertainty and noise. Early on, in the latter part of the 18th century, the flyball centrifugal governor enabled effective speed control of the steam turbine and thereby shares credit for the industrial revolution. Ever since, control has played a key role as an "enabling technology" in applications ranging from autopilots, navigation and telecommunications, to manufacturing, and power systems. The closing of the 20th century saw a rapid development of the mathematics of systems, control and optimization with a focus placed on understanding the benefits and limitations of feedback.

INTRODUCTION

Control engineering or control systems engineering is the engineering discipline that applies control theory to design systems with desired behaviors. The practice uses sensors to measure the output performance of the device being controlled and those measurements can be used to give feedback to the input actuators that can make corrections toward desired performance. When a device is designed to perform without the need of human inputs for correction it is called automatic control (such as cruise control for regulating the speed of a car). Multi-disciplinary in nature, control systems engineering activities focus on implementation of control systems mainly derived by mathematical modeling of systems of a diverse range.

Control systems play a critical role in space flight

Modern day control engineering is a relatively new field of study that gained significant attention during the 20th century with the advancement of technology. It can be broadly defined or classified as practical application of control theory. Control engineering has an essential role in a wide range of control systems, from simple household washing machines to high-performance F-16 fighter aircraft. It seeks to understand physical systems, using mathematical modeling, in terms of inputs, outputs and various components with different behaviors, use control systems design tools to develop controllers for those systems and implement controllers in physical systems employing available technology. A system can be mechanical, electrical, fluid, chemical, financial and even biological, and the mathematical modeling, analysis and controller design uses control theory in one or many of the time, frequency and complex-s domains, depending on the nature of the design problem.

Automatic control systems were first developed over two thousand years ago. The first feedback control device on record is thought to be the ancient Ktesibios's water clock in Alexandria, Egypt around the third century B.C. It kept time by regulating the water level in a vessel and, therefore, the water flow from that vessel. This certainly was a successful device as water clocks of similar design were still being made in Baghdad when the Mongols captured the city in 1258 A.D. A variety of automatic devices have been used over the centuries to accomplish useful tasks or simply to just entertain. The latter includes the automata, popular in Europe in the 17th and 18th centuries, featuring dancing figures that would repeat the same task over and over again; these automata are examples of open-loop control. Milestones among feedback, or "closed-loop" automatic control devices, include the temperature regulator of a furnace attributed to Drebbel, circa 1620, and the centrifugal flyball governor used for regulating the speed of steam engines by James Watt in 1788.

In his 1868 paper "On Governors", James Clerk Maxwell was able to explain instabilities exhibited by the flyball governor using differential equations to describe the control system. This demonstrated the importance and usefulness of mathematical models and methods in understanding complex phenomena, and it signaled the beginning of mathematical control and systems theory. Elements of control theory had appeared earlier but not as dramatically and convincingly as in Maxwell's analysis.

Control theory made significant strides over the next century. New mathematical techniques, as well as advancements in electronic and computer technologies, made it possible to control significantly more complex dynamical systems than the original flyball governor could stabilize. New mathematical techniques included developments in optimal control in the 1950s and 1960s followed by progress in stochastic, robust, adaptive, nonlinear, and azid-based control methods in the 1970s and 1980s. Applications of control methodology have helped to make possible space travel and communication satellites, safer and more efficient aircraft, cleaner automobile engines, and cleaner and more efficient chemical processes.

Before it emerged as a unique discipline, control engineering was practiced as a part of mechanical engineering and control theory was studied as a part of electrical engineering since electrical circuits can often be easily described using control theory techniques. In the very first control relationships, a current output was represented by a voltage control input. However, not having adequate technology to implement electrical control systems, designers were left with the option of less efficient and slow responding mechanical systems. A very effective mechanical controller that is still widely used in some hydro plants is the governor. Later on, previous to modern power electronics, process control systems for industrial applications were devised by mechanical engineers using pneumatic and hydraulic control devices, many of which are still in use today.

Control Theory

There are two major divisions in control theory, namely, classical and modern, which have direct implications for the control engineering applications. The scope of classical control theory is limited to single-input and single-output (SISO) system design, except when analyzing for disturbance rejection using a second input. The system analysis is carried out in the time domain using differential equations, in the complex-s domain with the Laplace transform, or in the frequency domain by transforming from the complex-s domain. Many systems may be assumed to have a second order and single variable system response in the time domain. A controller designed using classical theory often requires on-site tuning due to incorrect design approximations. Yet, due to the easier physical implementation of classical controller designs as compared to systems designed using modern control theory, these controllers are preferred in most industrial applications. The most common controllers designed using classical control theory are PID controllers. A less common implementation may include either or both a Lead or Lag filter. The ultimate end goal is to meet requirements typically provided in the time-domain called the step response, or at times in the frequency domain called the open-loop response. The step response characteristics applied in a specification are typically percent overshoot, settling time, etc. The open-loop response characteristics applied in a specification are typically Gain and Phase margin and bandwidth. These characteristics may be evaluated through simulation including a dynamic model of the system under control coupled with the compensation model.

In contrast, modern control theory is carried out in the state space, and can deal with multiple-input and multiple-output (MIMO) systems. This overcomes the limitations of classical control theory in more sophisticated design problems, such as fighter aircraft control, with the limitation that no frequency domain analysis is possible. In modern design, a system is represented to the greatest advantage as a set of decoupled first order differential equations defined using state variables. Nonlinear, multivariable, adaptive and robust control theories come under this division. Matrix methods are significantly limited for MIMO systems where linear independence cannot be assured in the relationship between inputs and outputs. Being fairly new, modern control theory has many areas yet to be explored. Scholars like Rudolf E. Kalman and Aleksandr Lyapunov are well-known among the people who have shaped modern control theory.

Control Systems

Control engineering is the engineering discipline that focuses on the modeling of a diverse range of dynamic systems (e.g. mechanical systems) and the design of controllers that will cause these systems to behave in the desired manner. Although such controllers need not be electrical many are and hence control engineering is often viewed as a subfield of electrical engineering. However, the falling price of microprocessors is making the actual implementation of a control system essentially trivial. As a result, focus is shifting back to the mechanical and process engineering discipline, as intimate knowledge of the physical system being controlled is often desired.

Electrical circuits, digital signal processors and microcontrollers can all be used to implement control systems. Control engineering has a wide range of applications from the flight and propulsion systems of commercial airliners to the cruise control present in many modern automobiles.

In most of the cases, control engineers utilize feedback when designing control systems. This is often accomplished using a PID controller system. For example, in an automobile with cruise control the vehicle's speed is continuously monitored and fed back to the system, which adjusts the motor's torque accordingly. Where there is regular feedback, control theory can be used to determine how the system responds to such feedback. In practically all such systems stability is important and control theory can help ensure stability is achieved.

Although feedback is an important aspect of control engineering, control engineers may also work on the control of systems without feedback. This is known as open loop control. A classic example of open loop control is a washing machine that runs through a pre-determined cycle without the use of sensors.

Control Engineering Education

At many universities, control engineering courses are taught in electrical and electronic engineering, mechatronics engineering, mechanical engineering, and aerospace engineering. In others, control engineering is connected to computer science, as most control techniques today are implemented through computers, often as embedded systems (as in the automotive field). The field of control within chemical engineering is often known as process control.

It deals primarily with the control of variables in a chemical process in a plant. It is taught as part of the undergraduate curriculum of any chemical engineering program and employs many of the same principles in control engineering. Other engineering disciplines also overlap with control engineering as it can be applied to any system for which a suitable model can be derived. However, specialised control engineering departments do exist, for example, the Department of Automatic Control and Systems Engineering at the University of Sheffield and the Department of Systems Engineering at the United States Naval Academy. Control engineering has diversified applications that include science, finance management, and even human behavior. Students of control engineering may start with a linear control system course dealing with the time and complex-s domain, which requires a thorough background in elementary mathematics and Laplace transform, called classical control theory. In linear control, the student does frequency and time domain analysis. Digital control and nonlinear control courses require Z transformation and algebra respectively, and could be said to complete a basic control education.

Recent Advancement

Originally, control engineering was all about continuous systems. Development of computer control tools posed a requirement of discrete control system engineering because the communications between the computer-based digital controller and the physical system are governed by a computer clock. The equivalent to Laplace transform in the discrete domain is the Z-transform. Today, many of the control systems are computer controlled and they consist of both digital and analog components.

Therefore, at the design stage either digital components are mapped into the continuous domain and the design is carried out in the continuous domain, or analog components are mapped into discrete domain and design is carried out there. The first of these two methods is more commonly encountered in practice because many industrial systems have many continuous systems components, including mechanical, fluid, biological and analog electrical components, with a few digital controllers. Similarly, the design technique has progressed from paper-and-ruler based manual design to computer-aided design and now to computer-automated design or CAutoD which has been made possible by evolutionary computation. CAutoD can be applied not just to tuning a predefined control scheme, but also to controller structure optimisation, system identification and invention of novel control systems, based purely upon a performance requirement, independent of any specific control scheme. Resilient control systems extend the traditional focus of addressing only planned disturbances to frameworks and attempt to address multiple types of unexpected disturbance; in particular, adapting and transforming behaviors of the control system in response to malicious actors, abnormal failure modes, undesirable human action, etc.

ADVANCED PROCESS CONTROL

In control theory Advanced process control (APC) refers to a broad range of techniques and technologies implemented within industrial process control systems. Advanced process controls are usually deployed optionally and in addition to *basic* process controls. Basic process controls are designed and built with the process itself, to facilitate basic operation, control and automation requirements. Advanced process controls are typically added subsequently, often over the course of many years, to address particular performance or economic improvement opportunities in the process. Process control (basic and advanced) normally implies the process industries, which includes chemicals, petrochemicals, oil and mineral refining, food processing, pharmaceuticals, power generation, etc. These industries are characterized by continuous processes and fluid processing, as opposed to discrete parts manufacturing, such as automobile and electronics manufacturing. The term process automation is essentially synonymous with process control. Process controls (basic as well as advanced) are implemented within the process control system, which usually means a distributed control system (DCS), programmable logic controller (PLC), and/or a supervisory control computer. DCSs and PLCs are typically industrially hardened and fault-tolerant. Supervisory control computers are often not hardened or fault-tolerant, but they bring a higher level of computational capability to the control system, to host valuable, but not critical, advanced control applications. Advanced controls may reside in either the DCS or the supervisory computer, depending on the application. Basic controls reside in the DCS and its subsystems, including PLCs.

Types of Advanced Process Control

Following is a list of the best known types of advanced process control:

- Advanced regulatory control (ARC) refers to several proven advanced control techniques, such as feedforward, override or adaptive gain. ARC is also a catch-all term used to refer to any customized or non-simple technique that does not fall into any other category. ARCs are typically implemented using function blocks or custom programming capabilities at the DCS level. In some cases, ARCs reside at the supervisory control computer level.
- Multivariable Model predictive control (MPC) is a popular technology, usually deployed on a supervisory control computer, that identifies important independent and dependent process variables and the dynamic relationships (models) between them, and uses matrix-math based control and optimization algorithms, to control multiple variables simultaneously. One requirement of MPC is that the models must be linear (i.e. they must be repeatable at all points of the operating range of the controller). MPC has been a prominent part of APC ever since supervisory computers first brought the necessary computational capabilities to control systems in the 1980s.
- Inferential control: The concept behind inferentials is to calculate a stream property from readily available process measurements, such as temperature and pressure, that otherwise would require either an expensive and complicated online analyzer or periodic laboratory analysis. Inferentials can be utilized in place of actual online analyzers, whether for operator information, cascaded to base-layer process controllers, or multivariable controller CVs.
- Sequential control refers to dis-continuous time and event based automation sequences that occur within continuous processes. These may be implemented as a collection of time and logic function blocks, a custom algorithm, or using a formalized Sequential function chart methodology.

- Compressor control typically includes compressor anti-surge and performance control.
- Nonlinear MPC: Similar to Multivariable MPC in that it incorporates dynamic models and matrix-math based control; however, it does not have the requirement for model linearity. Nonlinear MPC is capable of accommodating processes with models that have varying process gains and dynamics (i.e. dead-times and lag times).

Related Technologies

The following technologies are related to APC and in some contexts can be considered part of APC, but are generally separate technologies having their own (or in need of their own) Wiki articles.

- Statistical process control (SPC), despite its name, is much more common in discrete parts manufacturing and batch process control than in continuous process control. In SPC, "process" refers to the work and quality control process, rather than continuous process control.
- Batch process control (see ANSI/ISA-88) is employed in non-continuous batch processes, such as many pharmaceuticals, chemicals, and foods.
- Simulation-based optimization incorporates dynamic or steady-state computer-based process simulation models to determine more optimal operating targets in real-time, i.e. on a periodic basis, ranging from hourly to daily. This is sometimes considered a part of APC, but in practice it is still an emerging technology and is more often part of MPO.
- Manufacturing planning and optimization (MPO) refers to ongoing business activity to arrive at optimal operating targets that are then implemented in the operating organization, either manually or in some cases automatically communicated to the process control system.
- Safety instrumented system refers to a system that is independent of the process control system, both physically and administratively, whose purpose is to assure basic safety of the process.

APC Business and Professionals

Those responsible for the design, implementation and maintenance of APC applications are often referred to as APC Engineers or Control Application Engineers. Usually their education is dependent upon the field of specialization. For example, in the process industries many APC Engineers have a chemical engineering background, combining process control and chemical processing expertise. Most large operating facilities, such as oil refineries, employ a number of control system specialists and professionals, ranging from field instrumentation, regulatory control system (DCS and PLC), advanced process control, and control system network and security. Depending on facility size and circumstances, these personnel may have responsibilities across multiple areas, or be dedicated to each area. There are also many process control service companies that can be hired for support and services in each area.

CONTROL RECONFIGURATION

Control reconfiguration is an active approach in control theory to achieve fault-tolerant control for dynamic systems. It is used when severe faults, such as actuator or sensor outages, cause a break-up of the control loop, which must be restructured to prevent failure at the system level. In addition to loop restructuring, the controller parameters must be adjusted to accommodate changed plant dynamics. Control reconfiguration is a building block toward increasing the dependability of systems under feedback control.

Reconfiguration Problem

Fault Modelling

The figure to the right shows a plant controlled by a controller in a standard control loop.

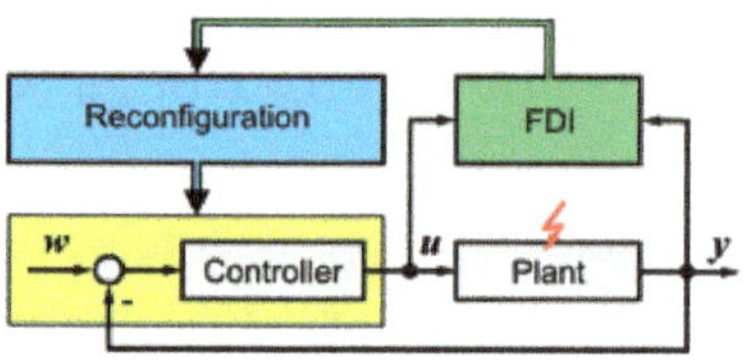

Schematic diagram of a typical active fault-tolerant control system. In the nominal, i.e. fault-free situation, the lower control loop operates to meet the control goals. The fault-detection (FDI) module monitors the closed-loop system to detect and isolate faults. The fault estimate is passed to the reconfiguration block, which modifies the control loop to reach the control goals in spite of the fault.

The nominal linear model of the plant is

$$\begin{cases} \dot{\mathbf{x}} &= \mathbf{Ax} + \mathbf{Bu} \\ \mathbf{y} &= \mathbf{Cx} \end{cases}$$

The plant subject to a fault (indicated by a red arrow in the figure) is modelled in general by

$$\begin{cases} \dot{\mathbf{x}}_f &= \mathbf{A}_f\mathbf{x}_f + \mathbf{B}_f\mathbf{u} \\ \mathbf{y}_f &= \mathbf{C}_f\mathbf{x}_f \end{cases}$$

where the subscript f indicates that the system is faulty. This approach models multiplicative faults by modified system matrices. Specifically, actuator faults are represented by the new input matrix B_f, sensor faults are represented by the output map C_f, and internal plant faults are represented by the system matrix A_f.

The upper part of the figure shows a supervisory loop consisting of *fault detection and isolation* (FDI) and *reconfiguration* which changes the loop by

- choosing new input and output signals from {u, y} to reach the control goal,
- changing the controller internals (including dynamic structure and parameters),
- adjusting the reference input w.

To this end, the vectors of inputs and outputs contain *all available signals*, not just those used by the controller in fault-free operation.

Alternative scenarios can model faults as an additive external signal f influencing the state derivatives and outputs as follows:

$$\begin{cases} \dot{\mathbf{x}}_f & = \mathbf{A}\mathbf{x}_f + \mathbf{B}\mathbf{u} + \mathbf{E}\mathbf{f} \\ \mathbf{y}_f & = \mathbf{C}_f\mathbf{x}_f + \mathbf{F}\mathbf{f} \end{cases}$$

Reconfiguration Goals

The goal of reconfiguration is to keep the reconfigured control-loop performance sufficient for preventing plant shutdown. The following goals are distinguished:

- Stabilization
- Equilibrium recovery
- Output trajectory recovery
- State trajectory recovery
- Transient time response recovery

Internal stability of the reconfigured closed loop is usually the minimum requirement. The equilibrium recovery goal (also referred to as weak goal) refers to the steady-state output equilibrium which the reconfigured loop reaches after a given constant input. This equilibrium must equal the nominal equilibrium under the same input (as time tends to infinity). This goal ensures steady-state reference tracking after reconfiguration. The output trajectory recovery goal (also referred to as strong goal) is even stricter. It requires that the dynamic response to an input must equal the nominal response at all times. Further restrictions are imposed by the state trajectory recovery goal, which requires that the state trajectory be restored to the nominal case by the reconfiguration under any input.

Usually a combination of goals is pursued in practice, such as the equilibrium-recovery goal with stability.

The question whether or not these or similar goals can be reached for specific faults is addressed by reconfigurability analysis.

Reconfiguration Approaches

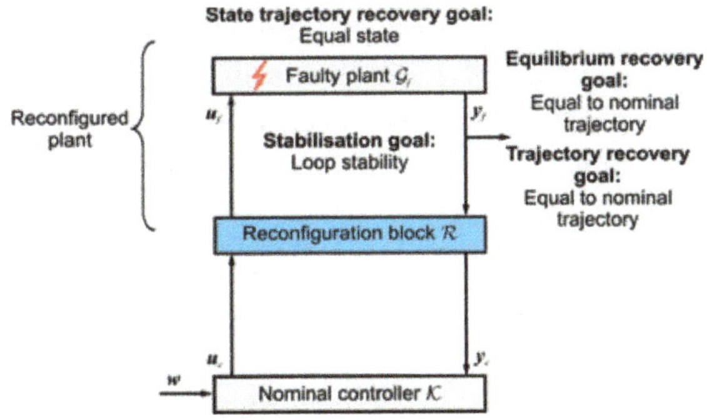

Fault hiding principle. A reconfiguration block is placed between faulty plant and nominal controller. The reconfuigured plant behaviour must match the nominal behaviour. Furthermore, the reconfiguration goals are pointed out.

Fault Hiding

This paradigm aims at keeping the nominal controller in the loop. To this end, a reconfiguration block can be placed between the faulty plant and the nominal controller. Together with the faulty plant, it forms the reconfigured plant. The reconfiguration block has to fulfill the requirement that the behaviour of the reconfigured plant matches the behaviour of the nominal, that is fault-free plant.

Linear model Following

In linear model following, a formal feature of the nominal closed loop is attempted to be recovered. In the classical pseudo-inverse method, the closed loop system matrix $\bar{\mathbf{A}} = \mathbf{A} - \mathbf{BK}$ of a state-feedback control structure is used. The new controller K_f is found to approximate $\bar{\mathbf{A}}$ in the sense of an induced matrix norm.

In perfect model following, a dynamic compensator is introduced to allow for the exact recovery of the complete loop behaviour under certain conditions.

In eigenstructure assignment, the nominal closed loop eigenvalues and eigenvectors (the eigenstructure) is recovered to the nominal case after a fault.

Optimisation-based Control Schemes

Optimisation control schemes include: linear-quadratic regulator design (LQR), model predictive control (MPC) and eigenstructure assignment methods.

Probabilistic Approaches

Some probabilistic approaches have been developed.

QUANTITATIVE FEEDBACK THEORY

In control theory, quantitative feedback theory (QFT), developed by Isaac Horowitz (Horowitz, 1963; Horowitz and Sidi, 1972), is a frequency domain technique utilising the Nichols chart (NC) in order to achieve a desired robust design over a specified region of plant uncertainty. Desired time-domain responses are translated into frequency domain tolerances, which lead to bounds (or constraints) on the loop transmission function. The design process is highly transparent, allowing a designer to see what trade-offs are necessary to achieve a desired performance level.

Plant Templates

Usually any system can be represented by its Transfer Function (Laplace in continuous time domain), after getting the model of a system.

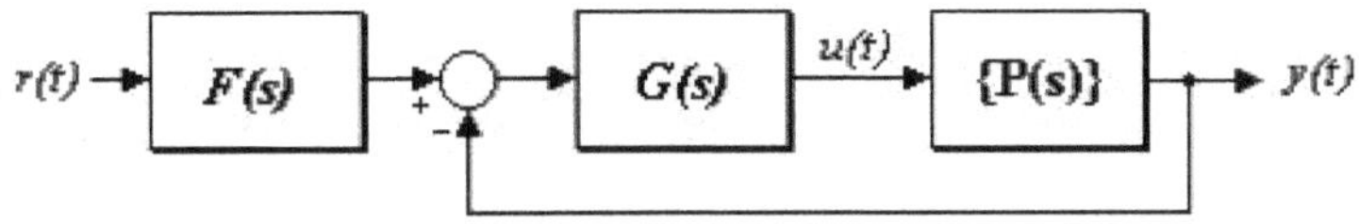

QFT Feedback System

As a result of experimental measurement, values of coefficients in the Transform Function have a range of uncertainty. Therefore, in QFT every parameter of this function is included into an interval of possible values, and the system may be represented by a family of plants rather than by a standalone expression.

$$\mathcal{P}(s) = \left\{ \frac{\prod_i (s+z_i)}{\prod_j (s+p_j)}, \forall z_i \in [z_{i,min}, z_{i,max}], p_j \in [p_{j,min}, p_{j,max}] \right\}$$

A frequency analysis is performed for a finite number of representative frequencies and a set of *templates* are obtained in the NC diagram which encloses the behaviour of the open loop system at each frequency.

Frequency Bounds

Usually system performance is described as robustness to **instability** (phase and gain margins), **rejection to input and output noise disturbances** and reference **tracking**. In the QFT design methodology these requirements on the system are represented as frequency constraints, conditions that the compensated system loop (controller and plant) could not break.

With these considerations and the selection of the same set of frequencies used for the templates, the frequency constraints for the behaviour of the system loop are computed and represented on the Nichols Chart (NC) as curves.

To achieve the problem requirements, a set of rules on the Open Loop Transfer Function, for the nominal plant $L_0(s) = G(s)P_0(s)$ may be found. That means the nominal loop is not allowed to have its frequency value below the constraint for the same frequency, and at high frequencies the loop should not cross the *Ultra High Frequency Boundary* (UHFB), which has an oval shape in the center of the NC.

Loop Shaping

The controller design is undertaken on the NC considering the frequency constraints and the *nominal loop* $L_0(s)$ of the system. At this point, the designer begins to introduce controller functions $(G(s))$ and tune their parameters, a process called Loop Shaping, until the best possible controller is reached without violation of the frequency constraints. The experience of the designer is an important factor in finding a satisfactory controller that not only complies with the frequency restrictions but with the possible realization, complexity, and quality. For this stage there currently exist different CAD (*Computer Aided Design*) packages to make the controller tuning easier.

Prefilter Design

Finally, the QFT design may be completed with a pre-filter $(F(s))$ design when it is required. In the case of tracking conditions a shaping on the Bode diagram may be used. Post design analysis is then performed to ensure the system response is satisfactory according with the problem requirements.

The QFT design methodology was originally developed for *Single-Input Single-Output* (SISO) and *Linear Time Invariant Systems* (LTI), with the design process being as described above. However, it has since been extended to weakly nonlinear systems, time varying systems, distributed parameter systems, multi-input multi-output (MIMO) systems (Horowitz, 1991), discrete systems (these using the Z-Transform as transfer function), and non minimum phase systems. The development of CAD tools has been an important, more recent development, which simplifies and automates much of the design procedure (Borghesani et al., 1994).

SLIDING MODE CONTROL

In control system, sliding mode control, or SMC, is a nonlinear control method that alters the dynamics of a nonlinear system by application of a discontinuous control signal that forces the system to "slide" along a cross-section of the system's normal behavior. The state-feedback control law is not a continuous function of time. Instead, it can switch from one continuous structure to another based on the current position in the state space. Hence, sliding mode control is a variable structure control method. The multiple control structures are designed so that trajectories always move toward an adjacent region with a different control structure, and so the ultimate trajectory will not exist entirely within one control structure. Instead, it will *slide* along the boundaries of the control structures. The motion of the system as it slides along these boundaries is called a *sliding mode* and the geometrical locus consisting of the boundaries is called the *sliding (hyper)surface*. In the context of modern control theory, any variable structure system, like a system under SMC, may be viewed as a special case of a hybrid dynamical system as the system both flows through a continuous state space but also moves through different discrete control modes.

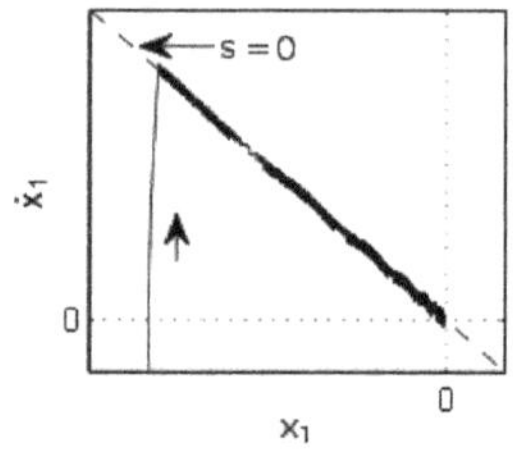

Figure 1: Phase plane trajectory of a system being stabilized by a sliding mode controller. After the initial reaching phase, the system states "slides" along the line S=0. The particular S=0 surface is chosen because it has desirable reduced-order dynamics when constrained to it. In this case, the $s = x_1 + \dot{x}_1 = 0$ surface corresponds to the first-order LTI system $\dot{x}_1 = -x_1$, which has an exponentially stable origin.

S=0, and the sliding mode along the surface commences after the finite time when system trajectories have reached the surface. In the theoretical description of sliding modes, the system stays confined to the sliding surface and need only be viewed as sliding along the surface. However, real implementations of sliding mode control approximate this theoretical behavior with a high-frequency and generally non-deterministic switching control signal that causes the system to "chatter" in a tight neighborhood of the sliding surface. In fact, although the system is nonlinear in general, the idealized (i.e., non-chattering) behavior of the system in Figure 1 when confined to the S=0 surface is an LTI system with an exponentially stable origin.

Intuitively, sliding mode control uses practically infinite gain to force the trajectories of a dynamic system to slide along the restricted sliding mode subspace. Trajectories from this reduced-order sliding mode have desirable properties (e.g., the system naturally slides along it until it comes to rest at a desired equilibrium). The main strength of sliding mode control is its robustness. Because the control can be as simple as a switching between two states, it need not be precise and will not be sensitive to parameter variations that enter into the control channel. Additionally, because the control law is not a continuous function, the sliding mode can be reached in *finite* time (i.e., better than asymptotic behavior). Under certain common conditions, optimality requires the use of bang–bang control; hence, sliding mode control describes the optimal controller for a broad set of dynamic systems.

One application of sliding mode controller is the control of electric drives operated by switching power converters. Because of the discontinuous operating mode of those converters, a discontinuous sliding mode controller is a natural implementation choice over continuous controllers that may need to be applied by means of pulse-width modulation or a similar technique[nb 1] of applying a continuous signal to an output that can only take discrete states. Sliding mode control has many applications in robotics. In particular, this control algorithm has been used for tracking control of unmanned surface vessels in simulated rough seas with high degree of success.

Sliding mode control must be applied with more care than other forms of nonlinear control that have more moderate control action. In particular, because actuators have delays and other imperfections, the hard sliding-mode-control action can lead to chatter, energy loss, plant damage, and excitation of unmodeled dynamics. Continuous control design methods are not as susceptible to these problems and can be made to mimic sliding-mode controllers.

Control Scheme

Consider a nonlinear dynamical system described by

$$\dot{\mathbf{x}}(t) = f(\mathbf{x}, t) + B(\mathbf{x}, t)\,\mathbf{u}(t)$$

where

$$\mathbf{x}(t) \triangleq \begin{bmatrix} x_1(t) \\ x_2(t) \\ \vdots \\ x_{n-1}(t) \\ x_n(t) \end{bmatrix} \in \mathbb{R}^n$$

is an {\displaystyle n}-dimensional state vector and

$$\mathbf{u}(t) \triangleq \begin{bmatrix} u_1(t) \\ u_2(t) \\ \vdots \\ u_{m-1}(t) \\ u_m(t) \end{bmatrix} \in \mathbb{R}^m$$

is an m-dimensional input vector that will be used for state feedback. The functions $f : \mathbb{R}^n \times \mathbb{R} \mapsto \mathbb{R}^n$ and {\displaystyle $B : \mathbb{R}^n \times \mathbb{R} \mapsto \mathbb{R}^{n \times m}$ are assumed to be continuous and sufficiently smooth so that the Picard–Lindelöf theorem can be used to guarantee that solution $\mathbf{x}(t)$ to Equation (1) exists and is unique. A common task is to design a state-feedback control law $\mathbf{u}(\mathbf{x}(t))$ (i.e., a mapping from current state x(t) at time t to the input u) to stabilize the dynamical system in Equation (1) around the origin $\mathbf{x} = [0, 0, \ldots, 0]^{\mathrm{T}}$. That is, under the control law, whenever the system is started away from the origin, it will return to it. For example, the component x_1 of the state vector X may represent the difference some output is away from a known signal (e.g., a desirable sinusoidal signal); if the control x_1 quickly returns to $x_1 = 0$, then the output will track the desired sinusoid. In sliding-mode control, the designer knows that the system behaves desirably (e.g., it has a stable equilibrium) provided that it is constrained to a subspace of its configuration space. Sliding mode control forces the system trajectories into this subspace and then holds them there so that they slide along it. This reduced-order subspace is referred to as a *sliding (hyper)surface*, and when closed-loop feedback forces trajectories to slide along it, it is referred to as a *sliding mode* of the closed-loop system. Trajectories along this subspace can be likened to trajectories along eigenvectors (i.e., modes) of LTI systems; however, the sliding mode is enforced by creasing the vector field with high-gain feedback. Like a marble rolling along a crack, trajectories are confined to the sliding mode.

The sliding-mode control scheme involves

- Selection of a hypersurface or a manifold (i.e., the sliding surface) such that the system trajectory exhibits desirable behavior when confined to this manifold.
- Finding feedback gains so that the system trajectory intersects and stays on the manifold.

Because sliding mode control laws are not continuous, it has the ability to drive trajectories to the sliding mode in finite time (i.e., stability of the sliding surface is better than asymptotic). However, once the trajectories reach the sliding surface, the system takes on the character of the sliding mode (e.g., the origin $\mathbf{x} = \mathbf{0}$ may only have asymptotic stability on this surface).

The sliding-mode designer picks a *switching function* $\sigma : \mathbb{R}^n \mapsto \mathbb{R}^m$ that represents a kind of "distance" that the states x are away from a sliding surface.

- A state x that is outside of this sliding surface has $\sigma(\mathbf{x}) \neq 0$.
- A state that is on this sliding surface has $\sigma(\mathbf{x}) = 0$.

The sliding-mode-control law switches from one state to another based on the *sign* of this distance. So the sliding-mode control acts like a stiff pressure always pushing in the direction of the sliding mode where $\sigma(\mathbf{x}) = 0$. Desirable $\mathbf{x}(t)$ trajectories will approach the sliding surface, and because the control law is not continuous (i.e., it switches from one state to another as trajectories move across this surface), the surface is reached in finite time. Once a trajectory reaches the surface, it will slide along it and may, for example, move toward the x=0 origin. So the switching function is like a topographic map with a contour of constant height along which trajectories are forced to move.

The sliding (hyper)surface is of dimension $n \times m$ is the number of states in u is the number of input signals (i.e., control signals) in u. For each control index $1 \leq k \leq m$, there is an $n \times 1$ sliding surface given by

$$\{\mathbf{x} \in \mathbb{R}^n : \sigma_k(\mathbf{x}) = 0\}$$

The vital part of SMC design is to choose a control law so that the sliding mode (i.e., this surface given by $\sigma(\mathbf{x}) = \mathbf{0}$)) exists and is reachable along system trajectories. The principle of sliding mode control is to forcibly constrain the system, by suitable control strategy, to stay on the sliding surface on which the system will exhibit desirable features. When the system is constrained by the sliding control to stay on the sliding surface, the system dynamics are governed by reduced-order system obtained from Equation (2).

To force the system states $\sigma(\mathbf{x}) = \mathbf{0}$to satisfy , one must:

- Ensure that the system is capable of reaching $\sigma(\mathbf{x}) = \mathbf{0}$ from any initial condition
- Having reached $\sigma(\mathbf{x}) = \mathbf{0}$, the control action is capable of maintaining the system at $\sigma(\mathbf{x}) = \mathbf{0}$

Existence of Closed-loop Solutions

Note that because the control law is not continuous, it is certainly not locally Lipschitz continuous, and so existence and uniqueness of solutions to the closed-loop system is *not* guaranteed by the Picard–Lindelöf theorem. Thus the solutions are to be understood in the Filippov sense. Roughly speaking, the resulting closed-loop system moving along $\sigma(\mathbf{x}) = \mathbf{0}$is approximated by the smooth dynamics $\sigma(\mathbf{x}) = \mathbf{0}$; however, this smooth behavior may not be truly realizable. Similarly, high-speed pulse-width modulation or delta-sigma modulation produces outputs that only assume two states, but the effective output swings through a continuous range of motion. These complications can be avoided by using a different nonlinear control design method that produces a continuous controller. In some cases, sliding-mode control designs can be approximated by other continuous control designs.

Theoretical Foundation

The following theorems form the foundation of variable structure control.

Theorem 1: Existence of Sliding Mode

Consider a Lyapunov function candidate

$$V(\sigma(\mathbf{x})) = \frac{1}{2}\sigma^{\mathrm{T}}(\mathbf{x})\sigma(\mathbf{x}) = \frac{1}{2}\|\sigma(\mathbf{x})\|_2^2$$

where $\|\cdot\|$ is the Euclidean norm (i.e $\|\sigma(\mathbf{x})\|_2$ is the distance away from the sliding manifold where $\sigma(\mathbf{x})=\mathbf{0}$). For the system given by Equation (1) and the sliding surface given by Equation (2), a sufficient condition for the existence of a sliding mode is that

$$\underbrace{\overbrace{\sigma^{\mathrm{T}}}^{\frac{\partial V}{\partial \sigma}}\overbrace{\dot{\sigma}}^{\frac{\mathrm{d}\sigma}{\mathrm{d}t}}}_{\frac{\mathrm{d}V}{\mathrm{d}t}}<0 \qquad \left(\text{i.e., } \frac{\mathrm{d}V}{\mathrm{d}t}<0\right)$$

in a neighborhood of the surface given by $\sigma(\mathbf{x})=\mathbf{0}$.

Roughly speaking (i.e., for the scalar control case when $m=1$), to achieve $\sigma^{\mathrm{T}}\dot{\sigma}<0$, , the feedback control law $u(\mathbf{x})$ is picked so that have opposite signs. That is,

- $u(\mathbf{x})$ makes $\dot{\sigma}(\mathbf{x})$ negative when $\sigma(\mathbf{x})$ is positive.
- $u(\mathbf{x})$ makes $\dot{\sigma}(\mathbf{x})$ positive when $\sigma(\mathbf{x})$ is negative.

Note that

$$\dot{\sigma}=\frac{\partial\sigma}{\partial\mathbf{x}}\overbrace{\dot{\mathbf{x}}}^{\frac{\mathrm{d}\mathbf{x}}{\mathrm{d}t}}=\frac{\partial\sigma}{\partial\mathbf{x}}\overbrace{(f(\mathbf{x},t)+B(\mathbf{x},t)\mathbf{u})}^{\dot{\mathbf{x}}}$$

and so the feedback control law {\displaystyle \mathbf {u} (\mathbf {x})} has a direct impact on {\displaystyle {\dot {\sigma }}}.

Reachability: Attaining sliding manifold in finite time

To ensure that the sliding mode is σ(x) = 0 attained in finite time, d V/d t must be more strongly bounded away from zero. That is, if it vanishes too quickly, the attraction to the sliding mode will only be asymptotic. To ensure that the sliding mode is entered in finite time,

$$\frac{\mathrm{d}V}{\mathrm{d}t}\leq-\mu(\sqrt{V})^{\alpha}$$

Explanation by Comparison Lemma

This condition ensures that for the neighborhood of the sliding mode $V\in[0,1]$,

$$\frac{\mathrm{d}V}{\mathrm{d}t}\leq-\mu(\sqrt{V})^{\alpha}\leq-\mu\sqrt{V}.$$

So, for $V\in(0,1]$,

$$\frac{1}{\sqrt{V}}\frac{\mathrm{d}V}{\mathrm{d}t}\leq-\mu,$$

which, by the chain rule (i.e., $\mathrm{d}W/\mathrm{d}t$ with $W\triangleq 2\sqrt{V}$), means

$$\underbrace{D^{+}\Big(\underbrace{2\overbrace{\sqrt{V}}^{\propto\|\sigma\|_2}}_{W}\Big)}_{D^{+}W \triangleq \text{Upper right-hand } \dot{W}} = \frac{1}{\sqrt{V}}\frac{\mathrm{d}V}{\mathrm{d}t} \le -\mu$$

where D^{+} is the upper right-hand derivative of $2\sqrt{V}$ and the symbol $z(t) = z_0 - \mu t$ denotes proportionality. So, by comparison to the curve $2\sqrt{V(t)} \le V_0 - \mu t$ which is represented by differential equation $\dot{z} = -\mu$ with initial condition $z(0) = z_0$, it must be the case that $2\sqrt{V(t)} \le V_0 - \mu t$ for all t. for all t. Moreover, because $\sqrt{V} \ge 0$, $\sqrt{V}$ must reach $\sqrt{V} = 0$ in finite time, which means that V must reach V=0 (i.e., the system enters the sliding mode) in finite time. Because $\sqrt{V}$ is proportional to the Euclidean norm $\|\cdot\|_2$ of the switching function σ, this result implies that the rate of approach to the sliding mode must be firmly bounded away from zero.

Consequences for Sliding Mode Control

In the context of sliding mode control, this condition means that

$$\underbrace{\overbrace{\sigma^{\mathrm{T}}}^{\frac{\partial V}{\partial \sigma}}\overbrace{\dot{\sigma}}^{\frac{\mathrm{d}\sigma}{\mathrm{d}t}}}_{\frac{\mathrm{d}V}{\mathrm{d}t}} \le -\mu(\overbrace{\|\sigma\|_2}^{\sqrt{V}})^{\alpha}$$

where $\|\cdot\|$ is the Euclidean norm. For the case when switching function σ is scalar valued, the sufficient condition becomes

$$\sigma\dot{\sigma} \le -\mu|\sigma|^{\alpha}.$$

Taking $\alpha = 1.$, the scalar sufficient condition becomes

$$\operatorname{sgn}(\sigma)\dot{\sigma} \le -\mu$$

which is equivalent to the condition that

$$\operatorname{sgn}(\sigma) \ne \operatorname{sgn}(\dot{\sigma}) \quad \text{and} \quad |\dot{\sigma}| \ge \mu > 0.$$

That is, the system should always be moving toward the switching surface σ = 0 and its speed $|\dot{\sigma}|$ toward the switching surface should have a non-zero lower bound. So, even though σ may become vanishingly small as σ (x) = 0 approaches the surface, $\dot{\sigma}$ must always be bounded firmly away from zero. To ensure this condition, sliding mode controllers are discontinuous across the σ = 0 manifold; they *switch* from one non-zero value to another as trajectories cross the manifold.

Automated Design Solutions

Although various theories exist for sliding mode control system design, there is a lack of a highly effective design methodology due to practical difficulties encountered in analytical and numerical methods. A reusable computing paradigm such as a genetic algorithm can, however, be utilized to transform a ‹unsolvable problem' of optimal design into a practically solvable ‹non-deterministic polynomial problem'. This results in computer-automated designs for sliding model control.

Sliding mode Observer

Sliding mode control can be used in the design of state observers. These non-linear high-gain observers have the ability to bring coordinates of the estimator error dynamics to zero in finite time. Additionally, switched-mode observers have attractive measurement noise resilience that is similar to a Kalman filter. For simplicity, the example here uses a traditional sliding mode modification of a Luenberger observer for an LTI system. In these sliding mode observers, the order of the observer dynamics are reduced by one when the system enters the sliding mode. In this particular example, the estimator error for a single estimated state is brought to zero in finite time, and after that time the other estimator errors decay exponentially to zero. However, as first described by Drakunov, a sliding mode observer for non-linear systems can be built that brings the estimation error for all estimated states to zero in a finite (and arbitrarily small) time.

Here, consider the LTI system

$$\begin{cases} \dot{\mathbf{x}} = A\mathbf{x} + B\mathbf{u} \\ y = \begin{bmatrix} 1 & 0 & 0 & \cdots & \end{bmatrix} \mathbf{x} = x_1 \end{cases}$$

where state vector $\mathbf{x} \triangleq (x_1, x_2, \ldots, x_n) \in \mathbb{R}^n$, $\mathbf{u} \triangleq (u_1, u_2, \ldots, u_r) \in \mathbb{R}^r$ is a vector of inputs, and output y is a scalar equal to the first state of the x state vector. Let

$$A \triangleq \begin{bmatrix} a_{11} & A_{12} \\ A_{21} & A_{22} \end{bmatrix}$$

where

- a_{11} is a scalar representing the influence of the first state x_1 on itself,
- $A_{21} \in \mathbb{R}^{(n-1)}$ is a column vector representing the influence of the other states on the first state,
- $A_{22} \in \mathbb{R}^{(n-1)\times(n-1)}$ is a matrix representing the influence of the other states on themselves, and
- $A_{12} \in \mathbb{R}^{1\times(n-1)}$ is a row vector corresponding to the influence of the first state on the other states.

The goal is to design a high-gain state observer that estimates the state vector x using only information from the measurement $y = x_1$ Hence, let the vector $\hat{\mathbf{x}} = (\hat{x}_1, \hat{x}_2, \ldots, \hat{x}_n) \in \mathbb{R}^n$ states. The observer takes the form

$$\dot{\hat{\mathbf{x}}} = A\hat{\mathbf{x}} + B\mathbf{u} + Lv(\hat{x}_1 - x_1)$$

where $v : \mathbb{R} \mapsto \mathbb{R}$ is a nonlinear function of the error between estimated state $\hat{x}_1$ and the output $y = x_1$, and $L \in \mathbb{R}^n$ is an observer gain vector that serves a similar purpose as in the typical linear Luenberger observer. Likewise, let $L = \begin{bmatrix} -1 \\ L_2 \end{bmatrix}$

where $L_2 \in \mathbb{R}^{(n-1)}$ is a column vector. Additionally, let $\mathbf{e} = (e_1, e_2, \ldots, e_n) \in \mathbb{R}^n$ be the state estimator error. That is, $\mathbf{e} = \hat{\mathbf{x}} - \mathbf{x}$. The error dynamics are then

$$\begin{cases} \dot{\mathbf{e}} = \dot{\hat{\mathbf{x}}} - \dot{\mathbf{x}} \\ = A\hat{\mathbf{x}} + B\mathbf{u} + Lv(\hat{x}_1 - x_1) - A\mathbf{x} - B\mathbf{u} \\ = A(\hat{\mathbf{x}} - \mathbf{x}) + Lv(\hat{x}_1 - x_1) \\ = A\mathbf{e} + Lv(e_1) \end{cases}$$

where $e_1 = \hat{x}_1 - x_1$ is the estimator error for the first state estimate. The nonlinear control law v can be designed to enforce the sliding manifold

$$0 = \hat{x}_1 - x_1$$

so that estimate $\hat{x}_1$ tracks the real state x_1 after some finite time (i.e., $\hat{x}_1 = x_1$). Hence, the sliding mode control switching function

$$\sigma(\hat{x}_1, \hat{x}) \triangleq e_1 = \hat{x}_1 - x_1.$$

To attain the sliding manifold, $\dot{\sigma}$ and σ must always have opposite signs for essentially all (i.e., x $\sigma\dot{\sigma} < 0$ for essentially. However,

$$\dot{\sigma} = \dot{e}_1 = a_{11}e_1 + A_{12}\mathbf{e}_2 - v(e_1) = a_{11}e_1 + A_{12}\mathbf{e}_2 - v(\sigma)$$

where $\mathbf{e}_2 \triangleq (e_2, e_3, \ldots, e_n) \in \mathbb{R}^{(n-1)}$ is the collection of the estimator errors for all of the unmeasured states. To ensure that $\sigma\dot{\sigma} < 0$, let

$$v(\sigma) = M\,\mathrm{sgn}(\sigma)$$

where

$$M > \max\{|a_{11}e_1 + A_{12}\mathbf{e}_2|\}.$$

That is, positive constant M must be greater than a scaled version of the maximum possible estimator errors for the system (i.e., the initial errors, which are assumed to be bounded so that M can be picked large enough; al). If M is sufficiently large, it can be assumed that the system achieves $e_1 = 0$ (i.e., $\hat{x}_1 = x_1$). Because e_1 is constant (i.e., 0) along this manifold, $\dot{e}_1 = 0$ as well. Hence, the discontinuous control $v(\sigma)$ may be replaced with the equivalent continuous control v_{eq} where

$$0 = \dot{\sigma} = a_{11}\overbrace{e_1}^{=0} + A_{12}\mathbf{e}_2 - \overbrace{v_{eq}}^{v(\sigma)} = A_{12}\mathbf{e}_2 - v_{eq}.$$

So

$$\overbrace{v_{eq}}^{\text{scalar}} = \overbrace{A_{12}}^{1\times(n-1)\text{ vector}} \overbrace{\mathbf{e}_2}^{(n-1)\times 1\text{ vector}}.$$

This equivalent control v_{eq} represents the contribution from the other (n-1) states to the trajectory of the output state x_1. In particular, the row A_{12} acts like an output vector for the error subsystem

$$\overbrace{\begin{bmatrix} \dot{e}_2 \\ \dot{e}_3 \\ \vdots \\ \dot{e}_n \end{bmatrix}}^{\dot{\mathbf{e}}_2} = A_2 \overbrace{\begin{bmatrix} e_2 \\ e_3 \\ \vdots \\ e_n \end{bmatrix}}^{\mathbf{e}_2} + L_2 v(e_1) = A_2\mathbf{e}_2 + L_2 v_{eq} = A_2\mathbf{e}_2 + L_2 A_{12}\mathbf{e}_2 = (A_2 + L_2 A_{12})\mathbf{e}_2.$$

So, to ensure the estimator error e_2 for the unmeasured states converges to zero, the $(n-1)\times 1$ vector L_2 must be chosen so that the $(n-1)\times(n-1)$ matrix $(A_2 + L_2 A_{12})$ is Hurwitz (i.e., the real part of each of its eigenvalues must be negative). Hence, provided that it is observable, this e_2 system can be stabilized in exactly the same way as a typical linear state observer when A_{12} is viewed as the output matrix (i.e., "C"). That is, the v_{eq} equivalent control provides measurement information about the unmeasured states that can continually move their estimates asymptotically closer to them. Meanwhile, the discontinuous control $v = M\,\mathrm{sgn}(\hat{x}_1 - x)$ forces the estimate of the measured state to have zero error in finite time. Additionally, white zero-mean symmetric measurement noise (e.g., Gaussian noise) only affects the switching frequency of the control v, and hence the noise will have little effect on the equivalent sliding mode control v_{eq}. Hence, the sliding mode observer has Kalman filter–like features.

The final version of the observer is thus

$$\begin{cases} \dot{\hat{\mathbf{x}}} = A\hat{\mathbf{x}} + B\mathbf{u} + LM\,\mathrm{sgn}(\hat{x}_1 - x_1) \\ = A\hat{\mathbf{x}} + B\mathbf{u} + \begin{bmatrix} -1 \\ L_2 \end{bmatrix} M\,\mathrm{sgn}(\hat{x}_1 - x_1) \\ = A\hat{\mathbf{x}} + B\mathbf{u} + \begin{bmatrix} -M \\ L_2 M \end{bmatrix} \mathrm{sgn}(\hat{x}_1 - x_1) \\ = A\hat{\mathbf{x}} + \begin{bmatrix} B & \begin{bmatrix} -M \\ L_2 M \end{bmatrix} \end{bmatrix} \begin{bmatrix} \mathbf{u} \\ \mathrm{sgn}(\hat{x}_1 - x_1) \end{bmatrix} \\ = A_{\mathrm{obs}}\hat{\mathbf{x}} + B_{\mathrm{obs}}\mathbf{u}_{\mathrm{obs}} \end{cases}$$

where

- $A_{\mathrm{obs}} \triangleq A$,
- $B_{\mathrm{obs}} \triangleq \begin{bmatrix} B & \begin{bmatrix} -M \\ L_2 M \end{bmatrix} \end{bmatrix}$, and
- $u_{\mathrm{obs}} \triangleq \begin{bmatrix} \mathbf{u} \\ \mathrm{sgn}(\hat{x}_1 - x_1) \end{bmatrix}$.

That is, by augmenting the control vector u with the switching function $\mathrm{sgn}(\hat{x}_1 - x_1)$, the sliding mode observer can be implemented as an LTI system. That is, the discontinuous signal $\mathrm{sgn}(\hat{x}_1 - x_1)$ is viewed as a control *input* to the 2-input LTI system.

For simplicity, this example assumes that the sliding mode observer has access to a measurement of a single state (i.e., output $y = x_1$). However, a similar procedure can be used to design a sliding mode observer for a vector of weighted combinations of states (i.e., when output $\mathbf{y} = C\mathbf{x}$ uses a generic matrix C). In each case, the sliding mode will be the manifold where the estimated output $\hat{\mathbf{y}}$ follows the measured output y with zero error (i.e., the manifold where $\sigma(\mathbf{x}) \triangleq \hat{\mathbf{y}} - \mathbf{y} = \mathbf{0}$).

REFRENCES

1. "ACSE - The University of Sheffield". Retrieved 17 March 2015.
2. USNA Weapons and Systems Engineering http://www.usna.edu/WSE/. Retrieved 19 November 2014. Missing or empty |title= (help)
3. Tan, K.C. and Li, Y. (2001) Performance-based control system design automation via evolutionary computing. Engineering Applications of Artificial Intelligence, 14 (4). pp. 473-486. ISSN 0952-1976, http://eprints.gla.ac.uk/3807/
4. Li, Y., et al. (2004). CAutoCSD - Evolutionary search and optimisation enabled computer automated control system design. International Journal of Automation and Computing, 1(1). pp.76-88. ISSN 1751-8520, http://eprints.gla.ac.uk/3818/
5. C. G. Rieger, D. I. Gertman and M. A. McQueen, "Resilient control systems: Next generation design research," 2009 2nd Conference on Human System Interactions, Catania, 2009, pp. 632-636. http://ieeexplore.ieee.org/stamp/stamp.jsp?tp=&arnumber=5091051&isnumber=5090940
6. (Blanke et al. 2006)
7. (Patton 1997)
8. (Steffen 2005)
9. (Gao & Antsaklis 1991) (Staroswiecki 2005)
10. (Looze et al. 1985)(Lunze, Rowe-Serrano & Steffen 2003)(Esna Ashari, Khaki Sedigh & Yazdanpanah 2005)(Maciejowski & Jones 2003)
11. (Mahmoud, Jiang & Zhang 2003)
12. (Rauch 1994)
13. (Zhang & Jiang 2003)
14. (Blanke et al. 2006)
15. Horowitz, I., 1963, Synthesis of Feedback Systems, Academic Press, New York, 1963.
16. Horowitz, I., and Sidi, M., 1972, "Synthesis of feedback systems with large plant ignorance for prescribed time-domain tolerances," International Journal of Control, 16(2), pp. 287–309.
17. Horowitz, I., 1991, "Survey of Quantitative Feedback Theory (QFT)," International Journal of Control, 53(2), pp. 255–291.
18. Borghesani, C., Chait, Y., and Yaniv, O., 1994, Quantitative Feedback Theory Toolbox Users Guide, The Math Works Inc., Natick, MA.
19. Zolotas, A. (2005, June 8). QFT - Quantitative Feedback Theory. Connexions.
20. Zinober, A.S.I., ed. (1990). Deterministic control of uncertain systems. London: Peter Peregrinus Press. ISBN 978-0-86341-170-0.
21. Utkin, Vadim I. (1993). "Sliding Mode Control Design Principles and Applications to Electric Drives". IEEE Transactions on Industrial Electronics. IEEE. **40** *(1): 23–36.* doi:10.1109/41.184818.
22. ‹Autonomous Navigation and Obstacle Avoidance of Unmanned Vessels in Simulated

Rough Sea States - Villanova University"

23. Mahini; et al. (2013). "An experimental setup for autonomous operation of surface vessels in rough seas". Robotica. **31** *(5): 703–715.* doi:10.1017/s0263574712000720.

24. Khalil, H.K. *(2002).* Nonlinear Systems *(3rd* ed.). Upper Saddle River, NJ: Prentice Hall. ISBN 0-13-067389-7.

25. Filippov, A.F. (1988). Differential Equations with Discontinuous Right-hand Sides. Kluwer. ISBN 978-90-277-2699-5.

26. Perruquetti, W.; Barbot, J.P. (2002). Sliding Mode Control in Engineering. Marcel Dekker Hardcover. ISBN 0-8247-0671-4.

27. Li, Yun; et al. (1996). "Genetic algorithm automated approach to the design of sliding mode control systems" *(PDF). International Journal of Control.* **64** *(3): 721–739.* doi:10.1080/00207179608921865.

28. Utkin, Vadim; Guldner, Jürgen; Shi, Jingxin (1999). Sliding Mode Control in Electromechanical Systems. Philadelphia, PA: Taylor & Francis, Inc. ISBN 0-7484-0116-4.

29. Drakunov, S.V. (1983). "An adaptive quasioptimal filter with discontinuous parameters". Automation and Remote Control. **44** *(9): 1167–1175.*

30. Drakunov, S.V. (1992). Sliding-Mode Observers Based on Equivalent Control Method. Proceedings of the 31st IEEE Conference on Decision and Control (CDC),. pp. 2368–2370. ISBN 0-7803-0872-7

31. Antunes, Ricardo; Gonzalez, Vicente (March 3, 2015). "A Production Model for Construction: A Theoretical Framework". Buildings. **5** *(1): 209–228.* doi:10.3390/buildings5010209.

32. "Feedback and control systems› - JJ Di Steffano, AR Stubberud, IJ Williams. Schaums outline series, McGraw-Hill 1967

33. Mayr, Otto (1970). The Origins of Feedback Control. Clinton, MA USA: The Colonial Press, Inc.

34. Mayr, Otto (1969). The Origins of Feedback Control. Clinton, MA USA: The Colonial Press, Inc.

35. Bennett, Stuart (1992). A history of control engineering, 1930-1955. IET. p. p. 48. ISBN 978-0-86341-299-8.

36. Ang, K.H., Chong, G.C.Y., and Li, Y. (2005). PID control system analysis, design, and technology, IEEE Trans Control Systems Tech, 13(4), pp.559–576.

37. Donald M Wiberg. State space & linear systems. Schaum's outline series. McGraw Hill. ISBN 0-07-070096-6.

38. Terrell, William (1999). "Some fundamental control theory I: Controllability, observability, and duality —AND— Some fundamental control Theory II: Feedback linearization of single input nonlinear systems". Amer. Math. Monthly. **106**: 705–719 and 812–828. doi:10.2307/2589614.

39. Gu Shi; et al. (2015). "Controllability of structural brain networks (Article Number 8414)". Nature Communications (6). doi:10.1038/ncomms9414. Lay summary. Here we use tools from control and network theories to offer a mechanistic explanation for how the brain moves between cognitive states drawn from the network organization of

white matter microstructure.

40. Liu, Jie; Wilson Wang; Farid Golnaraghi; Eric Kubica (2010). "A novel fuzzy framework for nonlinear system control". Fuzzy Sets and Systems. **161** *(21): 2746–2759.* doi:10.1016/j.fss.2010.04.009.

41. Maxwell, J.C. (1868). "On Governors". Proceedings of the Royal Society of London. **16**: 270–283. doi:10.1098/rspl.1867.0055. JSTOR 112510.

42. Routh, E.J.; Fuller, A.T. (1975). Stability of motion. Taylor & Francis.

43. Routh, E.J. (1877). A Treatise on the Stability of a Given State of Motion, Particularly Steady Motion: Particularly Steady Motion. Macmillan and co.

44. Hurwitz, A. (1964). "On The Conditions Under Which An Equation Has Only Roots With Negative Real Parts". Selected Papers on Mathematical Trends in Control Theory.

45. Richard Bellman (1964) Control Theory, Scientific American 211(3):186–200

CHAPTER 6

Classical Control Theory

Chapter Contents

The concepts of systems signals and Automatic Control, (brief historical review, basic structure of feedback control, examples) - Mathematical concepts and tools for the study of continuous and discrete time signals and systems. (Laplace transform, z-transform, applications, block diagrams and signal flow graphs) - Classification of signals and systems. Continuous and discrete time signals and systems - Time invariance, linearity - Classical analysis of systems and control in the time and frequency domains - Linear time invariant single-input, single-output systems described by ordinary, linear diferential equations - Input output relation and the transfer function description of a linear time invariant system - Free forced and total response of systems in the time domain - Stability of linear time invariant systems and algebraic stability criteria - Routh test for stability - Frequency response of linear time invariant systems - Closed loop systems - Root locus - Nyquist stability Criterion - Stabilizability and Stabilization of systems via precompensation and output feedback - Synthesis of controlers and paraperisation of stabilising controlers.

INTRODUCTION

Classical control theory is a branch of control theory that deals with the behavior of dynamical systems with inputs, and how their behavior is modified by feedback, using the Laplace transform as a basic tool to model such systems.

The usual objective of control theory is to control a system, often called the plant, so its output follows a desired control signal, called the reference, which may be a fixed or changing value. To do this a controller is designed, which monitors the output and compares it with the reference. The difference between actual and desired output, called the error signal, is applied as feedback to the input of the system, to bring the actual output closer to the reference.

Classical control theory deals with linear time-invariant single-input single-output systems. The Laplace transform of the input and output signal of such systems can be calculated. The transfer function relates the Laplace transform of the input and the output.

LINEAR TIME-INVARIANT THEORY

Linear time-invariant theory, commonly known as LTI system theory, comes from applied mathematics and has direct applications in NMR spectroscopy, seismology, circuits, signal processing, control theory, and other technical areas. It investigates the response of a linear and time-invariant system to an arbitrary input signal. Trajectories of these systems are commonly measured and tracked as they move through time (e.g., an acoustic waveform), but in applications like image processing and field theory, the LTI systems also have trajectories in spatial dimensions. Thus, these systems are also called linear translation-invariant to give the theory the most general reach. In the case of generic discrete-time (i.e., sampled) systems, linear shift-invariant is the corresponding term. A good example of LTI systems are electrical circuits that can be made up of resistors, capacitors, and inductors.

Overview

Linearity means that the relationship between the input and the output of the system is a linear map: If input x_1 (t) produces response y_1 (t) and input x_2 (t) produces response y_2 (t) then the *scaled* and *summed* input $a_1x_1(t) + a_2x_2(t)$ produces the scaled and summed response $a_1y_1(t) + a_2y_2(t)$ where a_1 and a_2 are real scalars. It follows that this can be extended to an arbitrary number of terms, and so for real numbers $c_1, c_2, \ldots, c_k$. In particular,

$$\text{Input } \sum_k c_k\, x_k(t) \text{ produces output } \sum_k c_k\, y_k(t).$$

$$\text{Input } \int_{-\infty}^{\infty} c_\omega\, x_\omega(t)\, \mathrm{d}\omega \text{ produces output } \int_{-\infty}^{\infty} c_\omega\, y_\omega(t)\, \mathrm{d}\omega$$

where cω and xω are scalars and inputs that vary over a continuum indexed by ω . Thus if an input function can be represented by a continuum of input functions, combined "linearly", as shown, then the corresponding output function can be represented by the corresponding continuum of output functions, *scaled* and *summed* in the same way.

- *Time invariance* means that whether we apply an input to the system now or *T* seconds from now, the output will be identical except for a time delay of *T* seconds. That is, if the output due to input $x(t)$ is $y(t)$, then the output due to input $x(t-T)$ is $y(t-T)$ Hence, the system is time invariant because the output does not depend on the particular time the input is applied.

- The fundamental result in LTI system theory is that any LTI system can be characterized entirely by a single function called the system's impulse response. The output of the system is simply the convolution of the input to the system with the system's impulse response. This method of analysis is often called the *time domain* point-of-view. The same result is true of discrete-time linear shift-invariant systems in which signals are discrete-time samples, and convolution is defined on sequences.

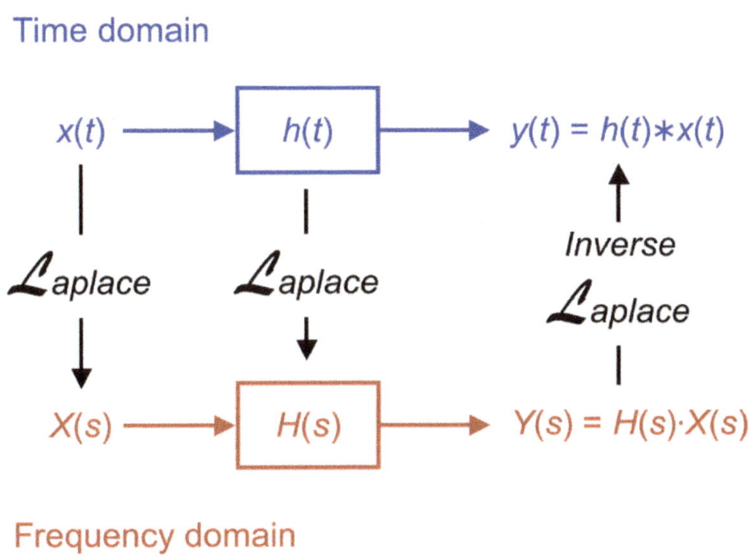

Relationship between the time domain and the frequency domain

Equivalently, any LTI system can be characterized in the *frequency domain* by the system's transfer function, which is the Laplace transform of the system's impulse response (or Z transform in the case of discrete-time systems). As a result of the properties of these transforms, the output of the system in the frequency domain is the product of the transfer function and the transform of the input. In other words, convolution in the time domain is equivalent to multiplication in the frequency domain. For all LTI systems, the eigenfunctions, and the basis functions of the transforms, are complex exponentials. This is, if the input to a system is the complex waveform Ae^{st} for some complex amplitude A and complex frequency s, Be^{st} for some new complex amplitude B. The ratio B/A is the transfer function at frequency s. the output will be some complex constant times the input, say

Since sinusoids are a sum of complex exponentials with complex-conjugate frequencies, if the input to the system is a sinusoid, then the output of the system will also be a sinusoid, perhaps with a different amplitude and a different phase, but always with the same frequency upon reaching steady-state. LTI systems cannot produce frequency components that are not in the input. LTI system theory is good at describing many important systems. Most LTI systems are considered "easy" to analyze, at least compared to the time-varying and/or nonlinear case. Any system that can be modeled as a linear homogeneous differential equation with constant coefficients is an LTI system. Examples of such systems are electrical circuits made up of resistors, inductors, and capacitors (RLC circuits). Ideal spring–mass–damper systems are also LTI systems, and are mathematically equivalent to RLC circuits.

Most LTI system concepts are similar between the continuous-time and discrete-time (linear shift-invariant) cases. In image processing, the time variable is replaced with two space variables, and the notion of time invariance is replaced by two-dimensional shift invariance. When analyzing filter banks and MIMO systems, it is often useful to consider vectors of signals.

A linear system that is not time-invariant can be solved using other approaches such as the Green function method. The same method must be used when the initial conditions of the problem are not null.

Continuous-Time Systems

Impulse Response and Convolution

The behavior of a linear, continuous-time, time-invariant system with input signal x(t) and output signal y(t) is described by the convolution integral:

$$\begin{aligned} y(t) = x(t) * h(t) &\stackrel{\text{def}}{=} \int_{-\infty}^{\infty} x(t-\tau)\cdot h(\tau)\,d\tau \\ &= \int_{-\infty}^{\infty} x(\tau)\cdot h(t-\tau)\,d\tau, \end{aligned}$$

(using commutativity)

where $h(t)$ is the system's response to an impulse: $x(\tau) = \delta(\tau)$. $y(t)$ is therefore proportional to a weighted average of the input function $x(\tau)$ The weighting function is $h(-\tau)$, simply shifted by amount t. As t changes, the weighting function emphasizes different parts of the input function. When $h(\tau)$ is zero for all negative τ, $y(t)$ depends only on values of x prior to time t and the system is said to be causal.

To understand why the convolution produces the output of an LTI system, let the notation $\{x(u-\tau);\ u\}$ represent the function $x(u-\tau)$ with variable u and constant τ And let the shorter notation {x} represent $\{x(u);\ u\}$. Then a continuous-time system transforms an input function, {x} into an output function, {y}. And in general, every value of the output can depend on every value of the input. This concept is represented by:

$$y(t) \stackrel{\text{def}}{=} O_t\{x\},$$

where Qt is the transformation operator for time t. In a typical system, y(t) depends most heavily on the values of x that occurred near time t . Unless the transform itself changes with t, the output function is just constant, and the system is uninteresting.

For a linear system, O must satisfy **Eq.1** :

$$O_t\left\{\int_{-\infty}^{\infty} c_\tau\, x_\tau(u)\,\mathrm{d}\tau;\ u\right\} = \int_{-\infty}^{\infty} c_\tau\, \underbrace{y_\tau(t)}_{O_t\{x_\tau\}}\,\mathrm{d}\tau.$$

And the time-invariance requirement is:

$$\begin{aligned} O_t\{x(u-\tau);\ u\} &= y(t-\tau) \\ &\stackrel{\text{def}}{=} O_{t-\tau}\{x\}. \end{aligned}$$

In this notation, we can write the **impulse response** as $h(t) \stackrel{\text{def}}{=} O_t\{\delta(u);\ u\}.$

Similarly:

$$\begin{aligned} h(t-\tau) &\stackrel{\text{def}}{=} O_{t-\tau}\{\delta(u);\ u\} \\ &= O_t\{\delta(u-\tau);\ u\}. \end{aligned}$$

Substituting this result into the convolution integral:

$$\begin{aligned} x(t) * h(t) &= \int_{-\infty}^{\infty} x(\tau)\cdot h(t-\tau)\,\mathrm{d}\tau \\ &= \int_{-\infty}^{\infty} x(\tau)\cdot O_t\{\delta(u-\tau);\ u\}\,\mathrm{d}\tau, \end{aligned}$$

which has the form of the right side of **Eq.2** for the case $c_\tau = x(\tau)$ and $x_\tau(u) = \delta(u-\tau)$. **Eq.2** then allows this continuation:

$$\begin{aligned} x(t) * h(t) &= O_t\left\{\int_{-\infty}^{\infty} x(\tau)\cdot\delta(u-\tau)\,\mathrm{d}\tau;\ u\right\} \\ &= O_t\left\{x(u);\ u\right\} \\ &\stackrel{\text{def}}{=} y(t). \end{aligned}$$

In summary, the input function, $\{x\}$, can be represented by a continuum of time-shifted impulse functions, combined "linearly", as shown at ***Eq.1***. The system's linearity property allows the system's response to be represented by the corresponding continuum of impulse responses, combined in the same way. And the time-invariance property allows that combination to be represented by the convolution integral.

The mathematical operations above have a simple graphical simulation.

Exponentials as Eigenfunctions

An eigenfunction is a function for which the output of the operator is a scaled version of the same function. That is,

$$\mathcal{H}f = \lambda f,$$

where f is the eigenfunction and λ is the eigenvalue, a constant.

The exponential functions Ae^{st}, where $A, s \in \mathbb{C}$ are eigenfunctions of a linear, time-invariant operator. A simple proof illustrates this concept. Suppose the input is $x(t) = Ae^{st}$. he output of the system with impulse response $h(t)$ is then

$$\int_{-\infty}^{\infty} h(t-\tau)Ae^{s\tau}\, \mathrm{d}\tau$$

which, by the commutative property of convolution, is equivalent to

$$\overbrace{\int_{-\infty}^{\infty} h(\tau)\, Ae^{s(t-\tau)}\, \mathrm{d}\tau}^{\mathcal{H}f} = \int_{-\infty}^{\infty} h(\tau)\, Ae^{st}e^{-s\tau}\, \mathrm{d}\tau \qquad = Ae^{st}\int_{-\infty}^{\infty} h(\tau)\, e^{-s\tau}\, \mathrm{d}\tau$$

$$= \underbrace{\overbrace{Ae^{st}}^{f}}_{\text{Input}}\underbrace{\overbrace{H(s)}^{\lambda}}_{\text{Scalar}},$$

where the scalar

$$H(s) \stackrel{\text{def}}{=} \int_{-\infty}^{\infty} h(t)e^{-st}\, \mathrm{d}t$$

is dependent only on the parameter s.

So the system's response is a scaled version of the input. In particular, for any $A, s \in \mathbb{C}$, the system output is the product of the input Ae^{st} and the constant $H(s)$. Hence, Ae^{st} is an eigenfunction of an LTI system, and the corresponding eigenvalue is $H(s)$.

Direct Proof

It is also possible to directly derive complex exponentials as eigenfunctions of LTI systems

Let's set $v(t) = e^{i\omega t}$ some complex exponential and $v_a(t) = e^{i\omega(t+a)}$ a time-shifted version of it.

$H[v_a](t) = e^{i\omega a}H[v](t)$ by linearity with respect to the constant $e^{i\omega a}$.

$H[v_a](t) = H[v](t+a)$ by time invariance of H.

So $H[v](t+a) = e^{i\omega a}H[v](t)$. Setting $t = 0$ and renaming we get :

$$H[v](\tau) = e^{i\omega\tau}H[v](0)$$

i.e. that a complex exponential $e^{i\omega\tau}$ as input will give a complex exponential of same frequency as output.

Fourier and Laplace Transforms

The eigenfunction property of exponentials is very useful for both analysis and insight into LTI systems. The Laplace transform

$$H(s) \stackrel{\text{def}}{=} \mathcal{L}\{h(t)\} \stackrel{\text{def}}{=} \int_{-\infty}^{\infty} h(t)e^{-st}\,\mathrm{d}t$$

is exactly the way to get the eigenvalues from the impulse response. Of particular interest are pure sinusoids (i.e., exponential functions of the form $e^{j\omega t}$ where $\omega \in \mathbb{R}$ and $j \stackrel{\text{def}}{=} \sqrt{-1}$). These are generally called complex exponentials even though the argument is purely imaginary. The Fourier transform $H(j\omega) = \mathcal{F}\{h(t)\}$ gives the eigenvalues for pure complex sinusoids. Both of H(s) and $H(j\omega)$ are called the *system function*, *system response*, or *transfer function*.

The Laplace transform is usually used in the context of one-sided signals, i.e. signals that are zero for all values of *t* less than some value. Usually, this "start time" is set to zero, for convenience and without loss of generality, with the transform integral being taken from zero to infinity (the transform shown above with lower limit of integration of negative infinity is formally known as the bilateral Laplace transform).

The Fourier transform is used for analyzing systems that process signals that are infinite in extent, such as modulated sinusoids, even though it cannot be directly applied to input and output signals that are not square integrable. The Laplace transform actually works directly for these signals if they are zero before a start time, even if they are not square integrable, for stable systems. The Fourier transform is often applied to spectra of infinite signals via the Wiener–Khinchin theorem even when Fourier transforms of the signals do not exist.

Due to the convolution property of both of these transforms, the convolution that gives the output of the system can be transformed to a multiplication in the transform domain, given signals for which the transforms exist

$$y(t) = (h * x)(t) \stackrel{\text{def}}{=} \int_{-\infty}^{\infty} h(t-\tau)x(\tau)\,\mathrm{d}\tau \stackrel{\text{def}}{=} \mathcal{L}^{-1}\{H(s)X(s)\}.$$

Not only is it often easier to do the transforms, multiplication, and inverse transform than the original convolution, but one can also gain insight into the behavior of the system from the system response. One can look at the modulus of the system function $|H(s)|$ to see whether the input $\exp(st)$ is *passed* (let through) the system or *rejected* or *attenuated* by the system (not let through).

Examples

- A simple example of an LTI operator is the derivative.

- $\frac{\mathrm{d}}{\mathrm{d}t}(c_1x_1(t)+c_2x_2(t)) = c_1x_1'(t)+c_2x_2'(t)$ (i.e., it is linear)
- $\frac{\mathrm{d}}{\mathrm{d}t}x(t-\tau) = x'(t-\tau)$ (i.e., it is time invariant)

When the Laplace transform of the derivative is taken, it transforms to a simple multiplication by the Laplace variable *s*.

$$\mathcal{L}\left\{\frac{\mathrm{d}}{\mathrm{d}t}x(t)\right\} = sX(s)$$

That the derivative has such a simple Laplace transform partly explains the utility of the transform.

- Another simple LTI operator is an averaging operator

$$\mathcal{A}\{x(t)\} \stackrel{\text{def}}{=} \int_{t-a}^{t+a} x(\lambda)\,\mathrm{d}\lambda.$$

By the linearity of integration,

$$\begin{aligned}\mathcal{A}\{c_1x_1(t)+c_2x_2(t)\} &= \int_{t-a}^{t+a}(c_1x_1(\lambda)+c_2x_2(\lambda))\,\mathrm{d}\lambda \\ &= c_1\int_{t-a}^{t+a}x_1(\lambda)\,\mathrm{d}\lambda + c_2\int_{t-a}^{t+a}x_2(\lambda)\,\mathrm{d}\lambda \\ &= c_1\mathcal{A}\{x_1(t)\}+c_2\mathcal{A}\{x_2(t)\},\end{aligned}$$

it is linear. Additionally, because

$$\begin{aligned}\mathcal{A}\{x(t-\tau)\} &= \int_{t-a}^{t+a} x(\lambda-\tau)\,\mathrm{d}\lambda \\ &= \int_{(t-\tau)-a}^{(t-\tau)+a} x(\xi)\,\mathrm{d}\xi \\ &= \mathcal{A}\{x\}(t-\tau),\end{aligned}$$

it is time invariant. In fact, A can be written as a convolution with the boxcar function $\Pi(t)$ That is

$$\mathcal{A}\{x(t)\} = \int_{-\infty}^{\infty} \Pi\left(\frac{\lambda-t}{2a}\right) x(\lambda)\,\mathrm{d}\lambda,$$

where the boxcar function

$$\Pi(t) \stackrel{\text{def}}{=} \begin{cases} 1 & \text{if } |t| < \frac{1}{2}, \\ 0 & \text{if } |t| > \frac{1}{2}. \end{cases}$$

Important System Properties

Some of the most important properties of a system are causality and stability. Causality is a necessity if the independent variable is time, but not all systems have time as an independent variable. For example, a system that processes still images does not need to be causal. Non-causal systems can be built and can be useful in many circumstances. Even non-real systems can be built and are very useful in many contexts.

Causality

A system is causal if the output depends only on present and past, but not future inputs. A necessary and sufficient condition for causality is

$$h(t) = 0 \quad \forall t < 0,$$

where h (t) is the impulse response. It is not possible in general to determine causality from the Laplace transform, because the inverse transform is not unique. When a region of convergence is specified, then causality can be determined.

Stability

A system is **bounded-input, bounded-output stable** (BIBO stable) if, for every bounded input, the output is finite. Mathematically, if every input satisfying

$$\|x(t)\|_\infty < \infty$$

leads to an output satisfying

$$\|y(t)\|_\infty < \infty$$

(that is, a finite maximum absolute value of $x(t)$ implies a finite maximum absolute value of $y(t)$), then the system is stable. A necessary and sufficient condition is that $h(t)$, the impulse response, is in L^1

(has a finite L^1 norm):

$$\|h(t)\|_1 = \int_{-\infty}^{\infty} |h(t)| \, \mathrm{d}t < \infty.$$

In the frequency domain, the region of convergence must contain the imaginary axis $s = j\omega$. As an example, the ideal low-pass filter with impulse response equal to a sinc function is not BIBO stable, because the sinc function does not have a finite L^1 norm. Thus, for some bounded input, the output of the ideal low-pass filter is unbounded. In particular, if the input is zero for $t < 0$ and equal to a sinusoid at the cut-off frequency for $t > 0$, then the output will be unbounded for all times other than the zero crossings.

Discrete-Time Systems

Almost everything in continuous-time systems has a counterpart in discrete-time systems.

Discrete-Time systems from Continuous-Time Systems

In many contexts, a discrete time (DT) system is really part of a larger continuous time (CT) system. For example, a digital recording system takes an analog sound, digitizes it, possibly processes the digital signals, and plays back an analog sound for people to listen to.

Formally, the DT signals studied are almost always uniformly sampled versions of CT signals. If $x(t)$ is a CT signal, then an analog to digital converter will transform it to the DT signal:

$$x[n] \stackrel{\text{def}}{=} x(nT) \qquad \forall n \in \mathbb{Z},$$

where T is the sampling period. It is very important to limit the range of frequencies in the input signal for faithful representation in the DT signal, since then the sampling theorem guarantees that no information about the CT signal is lost. A DT signal can only contain a frequency range of $1/(2T)$; other frequencies are aliased to the same range.

Impulse Response and Convolution

Let $\{x[m-k];\ m\}$ represent the sequence
$\{x[m-k];\ \textbf{for all integer values of m}\}$.

And let the shorter notation $\{x\}$ represent $\{x[m];\ m\}$.

A discrete system transforms an input sequence, {x} into an output sequence {y}. In general, every element of the output can depend on every element of the input. Representing the transformation operator by O, we can write:

$$y[n] \stackrel{\text{def}}{=} O_n\{x\}.$$

Note that unless the transform itself changes with **n**, the output sequence is just constant, and the system is uninteresting. (Thus the subscript, **n**.) In a typical system, **y[n]** depends most heavily on the elements of **x** whose indices are near **n**.

For the special case of the Kronecker delta function, $x[m] = \delta[m]$, the output sequence is the impulse response:

$$h[n] \stackrel{\text{def}}{=} O_n\{\delta[m];\ m\}.$$

For a linear system, O must satisfy:

$$O_n\left\{\sum_{k=-\infty}^{\infty} c_k \cdot x_k[m];\ m\right\} = \sum_{k=-\infty}^{\infty} c_k \cdot O_n\{x_k\}.$$

And the time-invariance requirement is:

$$\begin{aligned} O_n\{x[m-k];\ m\} &= y[n-k] \\ &\stackrel{\text{def}}{=} O_{n-k}\{x\}. \end{aligned}$$

In such a system, the impulse response, $\{h\}$, characterizes the system completely. I.e., for any input sequence, the output sequence can be calculated in terms of the input and the impulse response. To see how that is done, consider the identity:

$$x[m] \equiv \sum_{k=-\infty}^{\infty} x[k] \cdot \delta[m-k],$$

which expresses $\{x\}$ in terms of a sum of weighted delta functions.

Therefore:

$$\begin{aligned} y[n] = O_n\{x\} &= O_n\left\{\sum_{k=-\infty}^{\infty} x[k] \cdot \delta[m-k];\ m\right\} \\ &= \sum_{k=-\infty}^{\infty} x[k] \cdot O_n\{\delta[m-k];\ m\}, \end{aligned}$$

where we have invoked **Eq.4** for the case $c_k = x[k]$ and $x_k[m] = \delta[m-k]$.

And because of **Eq.5**, we may write:

$$\begin{aligned} O_n\{\delta[m-k];\ m\} &= O_{n-k}\{\delta[m];\ m\} \\ &\stackrel{\text{def}}{=} h[n-k]. \end{aligned}$$

Therefore:

$$\begin{aligned} y[n] &= \sum_{k=-\infty}^{\infty} x[k] \cdot h[n-k] \\ &= \sum_{k=-\infty}^{\infty} x[n-k] \cdot h[k], \end{aligned}$$

which is the familiar discrete convolution formula. The operator O_n can therefore be interpreted as proportional to a weighted average of the function **x[k]**. The weighting function is **h[-k]**, simply shifted by amount **n**. As **n** changes, the weighting function emphasizes different parts of the input function. Equivalently, the system's response to an impulse at **n=0** is a "time" reversed copy of the unshifted weighting function. When **h[k]** is zero for all negative **k**, the system is said to be causal.

Exponentials as Eigenfunctions

An eigenfunction is a function for which the output of the operator is the same function, scaled by some constant. In symbols,

$$\mathcal{H}f = \lambda f,$$

where f is the eigenfunction and λ is the eigenvalue, a constant. The exponential functions $z^n = e^{sTn}$, where $n \in \mathbb{Z}$, are eigenfunctions of a linear, time-invariant operator. $T \in \mathbb{R}$ is the sampling interval, and $z = e^{sT}, \ z, s \in \mathbb{C}$. A simple proof illustrates this concept. Suppose the input is $x[n] = z^n$. The output of the system with impulse response h[n] is then

$$\sum_{m=-\infty}^{\infty} h[n-m]\, z^m$$

which is equivalent to the following by the commutative property of convolution

$$\sum_{m=-\infty}^{\infty} h[m]\, z^{(n-m)} = z^n \sum_{m=-\infty}^{\infty} h[m]\, z^{-m} = z^n H(z)$$

where

$$H(z) \stackrel{\text{def}}{=} \sum_{m=-\infty}^{\infty} h[m] z^{-m}$$

is dependent only on the parameter z.

So z^n is an eigenfunction of an LTI system because the system response is the same as the input times the constant $H(z)$.

SINGLE-INPUT SINGLE-OUTPUT SYSTEM

In control engineering, a single-input and single-output (SISO) system is a simple single variable control system with one input and one output. In radio it is the use of only one antenna both in the transmitter and receiver.

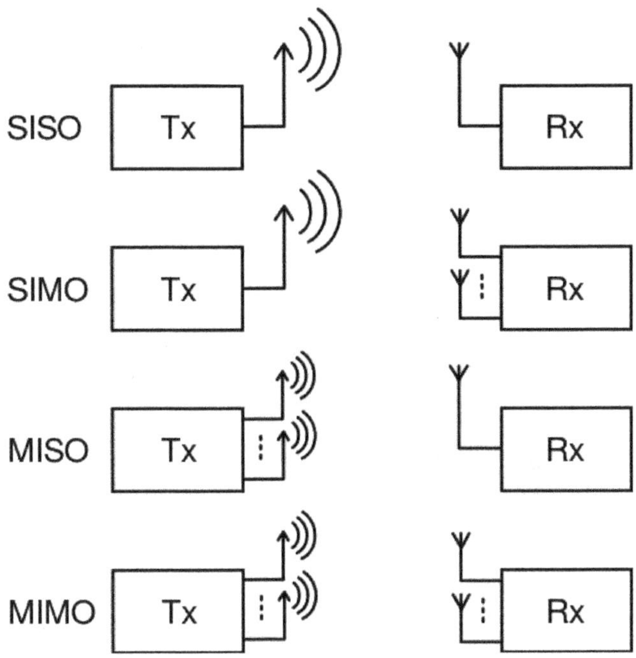

Understanding of SISO, SIMO, MISO and MIMO in radio systems

Details

SISO systems are typically less complex than multiple-input multiple-output system (MIMO) systems. Usually, it is also easier to make order of magnitude or trending predictions "on the fly" or "back of the envelope". MIMO systems have too many interactions for most of us to trace through them quickly, thoroughly, and effectively in our heads.

Frequency domain techniques for analysis and controller design dominate SISO control system theory. Bode plot, Nyquist stability criterion, Nichols plot, and root locus are the usual tools for SISO system analysis. Controllers can be designed through the polynomial design, root locus design methods to name just two of the more popular. Often SISO controllers will be PI, PID, or lead-lag.

OPEN-LOOP CONTROLLER

In an open-loop controller, also called a non-feedback controller, the control action from the controller is independent of the "process output", which is the process variable that is being controlled.

It does not use feedback to determine if its output has achieved the desired goal of the input or process "set point". An open-loop system cannot engage in machine learning and also cannot correct any errors that it could make. It will not compensate for disturbances in the process being controlled.

Open-Loop and Closed-Loop (Feedback) Control

Remote manipulator arms for working with radioactive materials – an open-loop mechanism, controlled by hand controls.

Fundamentally, there are two types of control loop: open loop control, and closed loop (feedback) control.

In open loop control, the control action from the controller is independent of the "process output" (or "controlled process variable"). A good example of this is a central heating boiler controlled only by a timer, so that heat is applied for a constant time, regardless of the temperature of the building. The control action is the switching on/off of the boiler. The process output is the building temperature.

In closed loop control, the control action from the controller is dependent on the process output. In the case of the boiler analogy this would include a thermostat to monitor the building temperature, and thereby feed back a signal to ensure the controller maintains the building at the temperature set on the thermostat. A closed loop controller therefore has a feedback loop which ensures the controller exerts a control action to give a process output the same as the "reference input" or "set point". For this reason, closed loop controllers are also called feedback controllers.

The definition of a closed loop control system according to the British Standard Institution is "a control system possessing monitoring feedback, the deviation signal formed as a result of this feedback being used to control the action of a final control element in such a way as to tend to reduce the deviation to zero." Likewise, a "'Feedback Control

System' is a system which tends to maintain a prescribed relationship of one system variable to another by comparing functions of these variables and using the difference as a means of control."

The advanced type of automation that revolutionized manufacturing, aircraft, communications and other industries, is feedback control, which is usually *continuous* and involves taking measurements using a sensor and making calculated adjustments to keep the measured variable within a set range. The theoretical basis of closed loop automation is the discipline of control theory.

Examples

An open-loop controller is often used in simple processes because of its simplicity and low cost, especially in systems where feedback is not critical. A typical example would be a conventional washing machine, for which the length of machine wash time is entirely dependent on the judgment and estimation of the human operator. Generally, to obtain a more accurate or more adaptive control, it is necessary to feed the output of the system back to the inputs of the controller. This type of system is called a closed-loop system.

For example, an irrigation sprinkler system, programmed to turn on at set times could be an example of an open-loop system if it does not measure soil moisture as a form of feedback. Even if rain is pouring down on the lawn, the sprinkler system would activate on schedule, wasting water.

Stepper motors are often used for open-loop control of position. A stepper motor rotates to one of a number of fixed positions, according to its internal construction. Sending a stream of electrical pulses to it causes it to rotate by exactly that many steps, hence the name. Such motors are often used, together with a simple initial datum sensor (a switch that is activated at the machine's home position), for the control of simple robotic machines or even the commonplace inkjet printer head. The drawback of open-loop control of steppers is that if the machine load is too high, or the motor attempts to move too quickly, then steps may be skipped. The controller has no means of detecting this and so the machine continues to run slightly out of adjustment, until reset. For this reason, more complex robots and machine tools instead use servomotors rather than stepper motors, which incorporate encoders and closed-loop controllers.

Open-loop control is useful for well-defined systems where the relationship between input and the resultant state can be modeled by a mathematical formula. For example, determining the voltage to be fed to an electric motor that drives a constant load, in order to achieve a desired speed would be a good application of open-loop control. If the load were not predictable, on the other hand, the motor's speed might vary as a function of the load as well as of the voltage, and an open-loop controller would therefore be insufficient to ensure repeatable control of the velocity.

An example of this is a conveyor system that is required to travel at a constant speed. For a constant voltage, the conveyor will move at a different speed depending on the load on the motor (represented here by the weight of objects on the conveyor). In order for the conveyor to run at a constant speed, the voltage of the motor must be adjusted depending on the load. In this case, a closed-loop control system would be necessary.

Linear and Nonlinear Control Theory

The field of control theory can be divided into two branches:

- *Linear control theory* – This applies to systems made of devices which obey the superposition principle, which means roughly that the output is proportional to the input. They are governed by linear differential equations. A major subclass is systems which in addition have parameters which do not change with time, called *linear time invariant* (LTI) systems. These systems are amenable to powerful frequency domain mathematical techniques of great generality, such as the Laplace transform, Fourier transform, Z transform, Bode plot, root locus, and Nyquist stability criterion. These lead to a description of the system using terms like bandwidth, frequency response, eigenvalues, gain, resonant frequencies, poles, and zeros, which give solutions for system response and design techniques for most systems of interest.
- *Nonlinear control theory* – This covers a wider class of systems that do not obey the superposition principle, and applies to more real-world systems, because all real control systems are nonlinear. These systems are often governed by nonlinear differential equations. The few mathematical techniques which have been developed to handle them are more difficult and much less general, often applying only to narrow categories of systems. These include limit cycle theory, Poincaré maps, Lyapunov stability theorem, and describing functions. Nonlinear systems are often analyzed using numerical methods on computers, for example by simulating their operation using a simulation language. If only solutions near a stable point are of interest, nonlinear systems can often be linearized by approximating them by a linear system using perturbation theory, and linear techniques can be used.

Analysis Techniques - Frequency Domain and Time Domain

Mathematical techniques for analyzing and designing control systems fall into two different categories:

- *Frequency domain* – In this type the values of the state variables, the mathematical variables representing the system's input, output and feedback are represented as functions of frequency. The input signal and the system's transfer function are converted from time functions to functions of frequency by a transform such as the Fourier transform, Laplace transform, or Z transform. The advantage of this technique is that it results in a simplification of the mathematics; the *differential equations* that represent the system are replaced by *algebraic equations* in the frequency domain which are much simpler to solve. However, frequency domain techniques can only be used with linear systems, as mentioned above.
- *Time-domain state space representation* – In this type the values of the state variables are represented as functions of time. With this model the system being

analyzed is represented by one or more differential equations. Since frequency domain techniques are limited to linear systems, time domain is widely used to analyze real-world nonlinear systems. Although these are more difficult to solve, modern computer simulation techniques such as simulation languages have made their analysis routine.

In contrast to the frequency domain analysis of the classical control theory, modern control theory utilizes the time-domain state space representation, a mathematical model of a physical system as a set of input, output and state variables related by first-order differential equations. To abstract from the number of inputs, outputs and states, the variables are expressed as vectors and the differential and algebraic equations are written in matrix form (the latter only being possible when the dynamical system is linear). The state space representation (also known as the 'time-domain approach') provides a convenient and compact way to model and analyze systems with multiple inputs and outputs. With inputs and outputs, we would otherwise have to write down Laplace transforms to encode all the information about a system. Unlike the frequency domain approach, the use of the state-space representation is not limited to systems with linear components and zero initial conditions. 'State space' refers to the space whose axes are the state variables. The state of the system can be represented as a point within that space.

System Interfacing - SISO & MIMO

Control systems can be divided into different categories depending on the number of inputs and outputs.

- Single-input single-output (SISO) – This is the simplest and most common type, in which one output is controlled by one control signal. Examples are the cruise control example above, or an audio system, in which the control input is the input audio signal and the output is the sound waves from the speaker.
- Multiple-input multiple-output (MIMO) – These are found in more complicated systems. For example, modern large telescopes such as the Keck and MMT have mirrors composed of many separate segments each controlled by an actuator. The shape of the entire mirror is constantly adjusted by a MIMO active optics control system using input from multiple sensors at the focal plane, to compensate for changes in the mirror shape due to thermal expansion, contraction, stresses as it is rotated and distortion of the wavefront due to turbulence in the atmosphere. Complicated systems such as nuclear reactors and human cells are simulated by computer as large MIMO control systems.

Topics in Control Theory

Stability

The *stability* of a general dynamical system with no input can be described with Lyapunov stability criteria.

- A linear system is called bounded-input bounded-output (BIBO) stable if its output will stay bounded for any bounded input.
- Stability for nonlinear systems that take an input is input-to-state stability (ISS), which combines Lyapunov stability and a notion similar to BIBO stability.

For simplicity, the following descriptions focus on continuous-time and discrete-time **linear systems**.

Mathematically, this means that for a causal linear system to be stable all of the poles of its transfer function must have negative-real values, i.e. the real part of each pole must be less than zero. Practically speaking, stability requires that the transfer function complex poles reside

- in the open left half of the complex plane for continuous time, when the Laplace transform is used to obtain the transfer function.
- inside the unit circle for discrete time, when the Z-transform is used.

The difference between the two cases is simply due to the traditional method of plotting continuous time versus discrete time transfer functions. The continuous Laplace transform is in Cartesian coordinates where the x axis is the real axis and the discrete Z-transform is in circular coordinates where the ρ axis is the real axis.

When the appropriate conditions above are satisfied a system is said to be asymptotically stable; the variables of an asymptotically stable control system always decrease from their initial value and do not show permanent oscillations. Permanent oscillations occur when a pole has a real part exactly equal to zero (in the continuous time case) or a modulus equal to one (in the discrete time case). If a simply stable system response neither decays nor grows over time, and has no oscillations, it is marginally stable; in this case the system transfer function has non-repeated poles at complex plane origin (i.e. their real and complex component is zero in the continuous time case). Oscillations are present when poles with real part equal to zero have an imaginary part not equal to zero.

If a system in question has an impulse response of

$$x[n] = 0.5^n u[n]$$

then the Z-transform (see this example), is given by

$$X(z) = \frac{1}{1 - 0.5z^{-1}}$$

which has a pole in $z = 0.5$ (zero imaginary part). This system is BIBO (asymptotically) stable since the pole is *inside* the unit circle.

However, if the impulse response was

$$x[n] = 1.5^n u[n]$$

then the Z-transform is

$$X(z) = \frac{1}{1 - 1.5z^{-1}}$$

which has a pole at z = 1.5 and is not BIBO stable since the pole has a modulus strictly greater than one.

Numerous tools exist for the analysis of the poles of a system. These include graphical systems like the root locus, Bode plots or the Nyquist plots.

Mechanical changes can make equipment (and control systems) more stable. Sailors add ballast to improve the stability of ships. Cruise ships use antiroll fins that extend transversely from the side of the ship for perhaps 30 feet (10 m) and are continuously rotated about their axes to develop forces that oppose the roll.

Controllability and Observability

Controllability and observability are main issues in the analysis of a system before deciding the best control strategy to be applied, or whether it is even possible to control or stabilize the system. Controllability is related to the possibility of forcing the system into a particular state by using an appropriate control signal. If a state is not controllable, then no signal will ever be able to control the state. If a state is not controllable, but its dynamics are stable, then the state is termed *stabilizable*. Observability instead is related to the possibility of *observing*, through output measurements, the state of a system. If a state is not observable, the controller will never be able to determine the behaviour of an unobservable state and hence cannot use it to stabilize the system. However, similar to the stabilizability condition above, if a state cannot be observed it might still be detectable.

From a geometrical point of view, looking at the states of each variable of the system to be controlled, every "bad" state of these variables must be controllable and observable to ensure a good behaviour in the closed-loop system. That is, if one of the eigenvalues of the system is not both controllable and observable, this part of the dynamics will remain untouched in the closed-loop system. If such an eigenvalue is not stable, the dynamics of this eigenvalue will be present in the closed-loop system which therefore will be unstable. Unobservable poles are not present in the transfer function realization of a state-space representation, which is why sometimes the latter is preferred in dynamical systems analysis.

Solutions to problems of uncontrollable or unobservable system include adding actuators and sensors.

Control Specification

Several different control strategies have been devised in the past years. These vary from extremely general ones (PID controller), to others devoted to very particular classes of systems (especially robotics or aircraft cruise control).

A control problem can have several specifications. Stability, of course, is always present. The controller must ensure that the closed-loop system is stable, regardless of the open-loop stability. A poor choice of controller can even worsen the stability of the open-loop system, which must normally be avoided. Sometimes it would be desired to obtain particular dynamics in the closed loop: i.e. that the poles have

$Re[\lambda] < -\overline{\lambda}$, where $\overline{\lambda}$ is a fixed value strictly greater than zero, instead of simply asking that $Re[\lambda] < 0$.

Another typical specification is the rejection of a step disturbance; including an integrator in the open-loop chain (i.e. directly before the system under control) easily achieves this. Other classes of disturbances need different types of sub-systems to be included.

Other "classical" control theory specifications regard the time-response of the closed-loop system. These include the rise time (the time needed by the control system to reach the desired value after a perturbation), peak overshoot (the highest value reached by the response before reaching the desired value) and others (settling time, quarter-decay). Frequency domain specifications are usually related to robustness (see after).

Modern performance assessments use some variation of integrated tracking error (IAE,ISA,CQI).

Model Identification and Robustness

A control system must always have some robustness property. A robust controller is such that its properties do not change much if applied to a system slightly different from the mathematical one used for its synthesis. This specification is important, as no real physical system truly behaves like the series of differential equations used to represent it mathematically. Typically a simpler mathematical model is chosen in order to simplify calculations, otherwise the true system dynamics can be so complicated that a complete model is impossible.

System Identification

The process of determining the equations that govern the model's dynamics is called system identification. This can be done off-line: for example, executing a series of measures from which to calculate an approximated mathematical model, typically its transfer function or matrix. Such identification from the output, however, cannot take account of unobservable dynamics. Sometimes the model is built directly starting from known physical equations, for example, in the case of a mass-spring-damper system we know that $m\ddot{x}(t) = -Kx(t) - \mathrm{B}\dot{x}(t)$. Even assuming that a "complete" model is used in designing the controller, all the parameters included in these equations (called "nominal parameters") are never known with absolute precision; the control system will have to behave correctly even when connected to physical system with true parameter values away from nominal. Some advanced control techniques include an "on-line" identification process (see later). The parameters of the model are calculated ("identified") while the controller itself is running. In this way, if a drastic variation of the parameters ensues, for example, if the robot's arm releases a weight, the controller will adjust itself consequently in order to ensure the correct performance.

Analysis

Analysis of the robustness of a SISO (single input single output) control system can be performed in the frequency domain, considering the system's transfer function and using Nyquist and Bode diagrams. Topics include gain and phase margin and amplitude margin. For MIMO (multi input multi output) and, in general, more complicated control systems one must consider the theoretical results devised for each control technique (see next section). I.e., if particular robustness qualities are needed, the engineer must shift his attention to a control technique by including them in its properties.

Constraints

A particular robustness issue is the requirement for a control system to perform properly in the presence of input and state constraints. In the physical world every signal is limited. It could happen that a controller will send control signals that cannot be followed by the physical system, for example, trying to rotate a valve at excessive speed. This can produce undesired behavior of the closed-loop system, or even damage or break actuators or other subsystems. Specific control techniques are available to solve the problem: model predictive control (see later), and anti-wind up systems. The latter consists of an additional control block that ensures that the control signal never exceeds a given threshold.

System Classifications

Linear Systems Control

For MIMO systems, pole placement can be performed mathematically using a state space representation of the open-loop system and calculating a feedback matrix assigning poles in the desired positions. In complicated systems this can require computer-assisted calculation capabilities, and cannot always ensure robustness. Furthermore, all system states are not in general measured and so observers must be included and incorporated in pole placement design.

Nonlinear Systems Control

Processes in industries like robotics and the aerospace industry typically have strong nonlinear dynamics. In control theory it is sometimes possible to linearize such classes of systems and apply linear techniques, but in many cases it can be necessary to devise from scratch theories permitting control of nonlinear systems. These, e.g., feedback linearization, backstepping, sliding mode control, trajectory linearization control normally take advantage of results based on Lyapunov's theory. Differential geometry has been widely used as a tool for generalizing well-known linear control concepts to the non-linear case, as well as showing the subtleties that make it a more challenging problem. Control theory has also been used to decipher the neural mechanism that directs cognitive states.

Decentralized Systems Control

When the system is controlled by multiple controllers, the problem is one of decentralized control. Decentralization is helpful in many ways, for instance, it helps control systems to operate over a larger geographical area. The agents in decentralized control systems can interact using communication channels and coordinate their actions.

Deterministic and Stochastic Systems Control

A stochastic control problem is one in which the evolution of the state variables is subjected to random shocks from outside the system. A deterministic control problem is not subject to external random shocks.

Main Control Strategies

Every control system must guarantee first the stability of the closed-loop behavior. For linear systems, this can be obtained by directly placing the poles. Non-linear control systems use specific theories (normally based on Aleksandr Lyapunov's Theory) to ensure stability without regard to the inner dynamics of the system. The possibility to fulfill different specifications varies from the model considered and the control strategy chosen.

List of the Main Control Techniques

- Adaptive control uses on-line identification of the process parameters, or modification of controller gains, thereby obtaining strong robustness properties. Adaptive controls were applied for the first time in the aerospace industry in the 1950s, and have found particular success in that field.
- A hierarchical control system is a type of control system in which a set of devices and governing software is arranged in a hierarchical tree. When the links in the tree are implemented by a computer network, then that hierarchical control system is also a form of networked control system.
- Intelligent control uses various AI computing approaches like neural networks, Bayesian probability, fuzzy logic, machine learning, evolutionary computation and genetic algorithms to control a dynamic system.
- Optimal control is a particular control technique in which the control signal optimizes a certain "cost index": for example, in the case of a satellite, the jet thrusts needed to bring it to desired trajectory that consume the least amount of fuel. Two optimal control design methods have been widely used in industrial applications, as it has been shown they can guarantee closed-loop stability. These are Model Predictive Control (MPC) and linear-quadratic-Gaussian control (LQG). The first can more explicitly take into account constraints on the signals in the system, which is an important feature in many industrial processes. However, the "optimal control" structure in MPC is only a means to achieve such a result, as it does not optimize a true performance index of the closed-loop control system. Together with PID controllers, MPC systems are the most widely used control technique in process control.
- Robust control deals explicitly with uncertainty in its approach to controller design. Controllers designed using *robust control* methods tend to be able to cope with small differences between the true system and the nominal model used for design. The early methods of Bode and others were fairly robust; the state-space methods invented in the 1960s and 1970s were sometimes found to

lack robustness. Examples of modern robust control techniques include H-infinity loop-shaping developed by Duncan McFarlane and Keith Glover of Cambridge University, United Kingdom and Sliding mode control (SMC) developed by Vadim Utkin. Robust methods aim to achieve robust performance and/or stability in the presence of small modeling errors.

- Stochastic control deals with control design with uncertainty in the model. In typical stochastic control problems, it is assumed that there exist random noise and disturbances in the model and the controller, and the control design must take into account these random deviations.
- Energy-shaping control view the plant and the controller as energy-transformation devices. The control strategy is formulated in terms of interconnection (in a power-preserving manner) in order to achieve a desired behavior.
- Self-organized criticality control may be defined as attempts to interfere in the processes by which the self-organized system dissipates energy.

Although control systems of various types date back to antiquity, a more formal analysis of the field began with a dynamics analysis of the centrifugal governor, conducted by the physicist James Clerk Maxwell in 1868, entitled *On Governors*. This described and analyzed the phenomenon of self-oscillation, in which lags in the system may lead to overcompensation and unstable behavior. This generated a flurry of interest in the topic, during which Maxwell's classmate, Edward John Routh, abstracted Maxwell's results for the general class of linear systems. Independently, Adolf Hurwitz analyzed system stability using differential equations in 1877, resulting in what is now known as the Routh–Hurwitz theorem.

A notable application of dynamic control was in the area of manned flight. The Wright brothers made their first successful test flights on December 17, 1903 and were distinguished by their ability to control their flights for substantial periods (more so than the ability to produce lift from an airfoil, which was known). Continuous, reliable control of the airplane was necessary for flights lasting longer than a few seconds.

By World War II, control theory was an important part of fire-control systems, guidance systems and electronics.

Sometimes, mechanical methods are used to improve the stability of systems. For example, ship stabilizers are fins mounted beneath the waterline and emerging laterally. In contemporary vessels, they may be gyroscopically controlled active fins, which have the capacity to change their angle of attack to counteract roll caused by wind or waves acting on the ship. The Sidewinder missile uses small control surfaces placed at the rear of the missile with spinning disks on their outer surfaces and these are known as rollerons. Airflow over the disks spins them to a high speed. If the missile starts to roll, the gyroscopic force of the disks drives the control surface into the airflow, cancelling the motion. Thus, the Sidewinder team replaced a potentially complex control system with a simple mechanical solution.

The Space Race also depended on accurate spacecraft control, and control theory has also seen an increasing use in fields such as economics.

People in Systems and Control

Many active and historical figures made significant contribution to control theory including

- Pierre-Simon Laplace (1749–1827) invented the Z-transform in his work on probability theory, now used to solve discrete-time control theory problems. The Z-transform is a discrete-time equivalent of the Laplace transform which is named after him.
- Alexander Lyapunov (1857–1918) in the 1890s marks the beginning of stability theory.
- Harold S. Black (1898–1983), invented the concept of negative feedback amplifiers in 1927. He managed to develop stable negative feedback amplifiers in the 1930s.
- Harry Nyquist (1889–1976) developed the Nyquist stability criterion for feedback systems in the 1930s.
- Richard Bellman (1920–1984) developed dynamic programming since the 1940s.
- Andrey Kolmogorov (1903–1987) co-developed the Wiener–Kolmogorov filter in 1941.
- Norbert Wiener (1894–1964) co-developed the Wiener–Kolmogorov filter and coined the term cybernetics in the 1940s.
- John R. Ragazzini (1912–1988) introduced digital control and the use of Z-transform in control theory (invented by Laplace) in the 1950s.
- Lev Pontryagin (1908–1988) introduced the maximum principle and the bang-bang principle.
- Pierre-Louis Lions (1956) developed viscosity solutions into stochastic control and optimal control methods.

PID CONTROLLER

A proportional–integral–derivative controller (PID controller) is a control loop feedback mechanism (controller) commonly used in industrial control systems. A PID controller continuously calculates an *error value* e(t) as the difference between a desired setpoint and a measured process variable and applies a correction based on proportional, integral, and derivative terms, (sometimes denoted *P*, *I*, and *D* respectively) which give their name to the controller type.

Fundamental Operation

A PID controller continuously calculates an error value $e(t)$ as the difference between a desired setpoint *and a measured process variable and applies a correction based on proportional, integral, and derivative terms. The controller attempts to minimize the error over time by adjustment of a control variable* $u(t)$, s such as the position of a control valve, a damper, or the power supplied to a heating element, to a new value determined by a weighted sum:

$$u(t) = K_p e(t) + K_i \int_0^t e(\tau)\, d\tau + K_d \frac{de(t)}{dt},$$

where K_p, K_i, and K_d all non-negative, denote the coefficients for the proportional, integral, and derivative terms, respectively (sometimes denoted *P*, *I*, and *D*). In this model:

- *P* accounts for present values of the error. For example, if the error is large and positive, the control output will also be large and positive.
- *I* accounts for past values of the error. For example, if the current output is not sufficiently strong, the integral of the error will accumulate over time, and the controller will respond by applying a stronger action.
- *D* accounts for possible future trends of the error, based on its current rate of change.

As a PID controller relies only on the measured process variable, not on knowledge of the underlying process, it is broadly applicable. By tuning the three parameters of the model, a PID controller can deal with specific process requirements. The response of the controller can be described in terms of its responsiveness to an error, the degree to which the system overshoots a setpoint, and the degree of any system oscillation. The use of the PID algorithm does not guarantee optimal control of the system or even its stability.

Some applications may require using only one or two terms to provide the appropriate system control. This is achieved by setting the other parameters to zero. A PID controller is called a PI, PD, P or I controller in the absence of the respective control actions. PI controllers are fairly common, since derivative action is sensitive to measurement noise, whereas the absence of an integral term may prevent the system from reaching its target value.

For discrete-time systems, the term PSD (proportional-summation-difference) is often used.

History and Applications

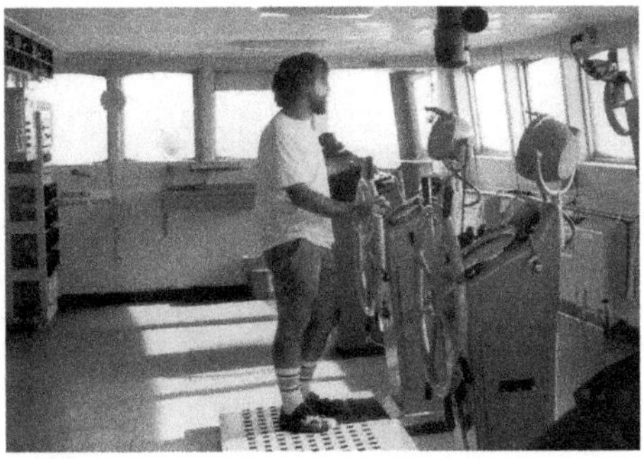

Early PID theory was developed by observing the actions of helmsmen in keeping a vessel on course in the face of varying influences such as wind and sea state.

Pneumatic PID (Three term) controller. The magnitude of the "three terms" P,I and D, are adjusted by the dials at the top.

Origins

PID controllers have their origins in 19th century speed governor design. The theoretical basis for the operation of governors was first described by James Clerk Maxwell in 1868 in his seminal paper 'On Governors', but it was not until 1922 that PID controllers were first developed using a theoretical analysis, by Russian American engineer Nicolas Minorsky (Minorsky 1922) for automatic ship steering. Minorsky was designing automatic steering systems for the US Navy and based his analysis on observations of a helmsman, noting the helmsman steered the ship based not only on the current course error, but also on past error, as well as the current rate of change; this was then given a mathematical treatment by Minorsky. His goal was stability, not general control, which simplified the problem significantly. While proportional control provides stability against small disturbances, it was insufficient for dealing with a steady disturbance, notably a stiff gale (due to steady-state error), which required adding the integral term. Finally, the derivative term was added to improve stability and control.

Trials were carried out on the USS *New Mexico*, with the controller controlling the *angular velocity* (not angle) of the rudder. PI control yielded sustained yaw (angular error) of ±2°. Adding the D element yielded a yaw error of ±1/6°, better than most helmsmen could achieve.

The Navy ultimately did not adopt the system, due to resistance by personnel. Similar work was carried out and published by several others in the 1930s.

One of the earliest examples of a PID-type controller was also developed by Elmer Sperry in 1911, though his work was intuitive rather than mathematically-based.

Industrial Controller Development

In the early history of automatic PID process control the first controllers were pneumatic devices, which used compressed air both for calculating the controller output, and for powering the process modulating device; such as a diaphragm operated valve. These controllers used nozzle and flapper high-gain pneumatic amplifiers and feedback to generate P, I and D control terms. They are simple low maintenance devices which operate well in a harsh industrial environment, and do not present an explosion risk in hazardous locations. They were the industry standard for many decades until the advent of discrete electronic controllers and distributed control systems.

With these controllers, a pneumatic industry signalling standard of 3-15 psi was established, which had an elevated zero to ensure devices were working within their linear characteristic and represented the control range of 0-100%. The metric equivalent for this is 0.2 - 1.0 bar.

In the 1950s when high gain electronic amplifiers became cheap and reliable, electronic controllers became popular, and 4-20 mA current loop signals were used which emulated the pneumatic standard. However field actuators still widely use the pneumatic standard because of the advantages of pneumatic motive power for control valves in process plant environments.

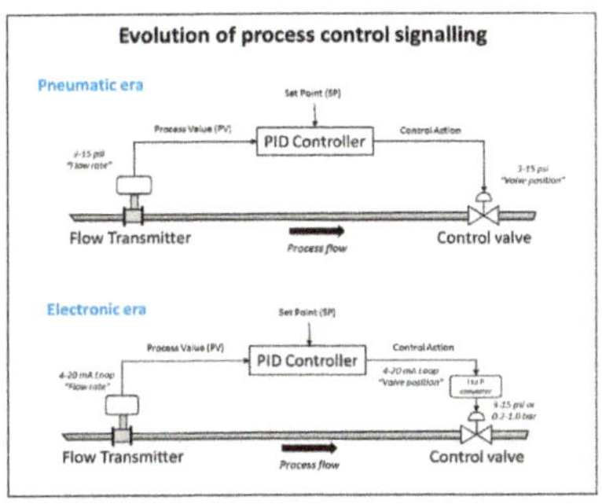

Showing the evolution of analogue control loop signalling from the pneumatic era to the electronic era.

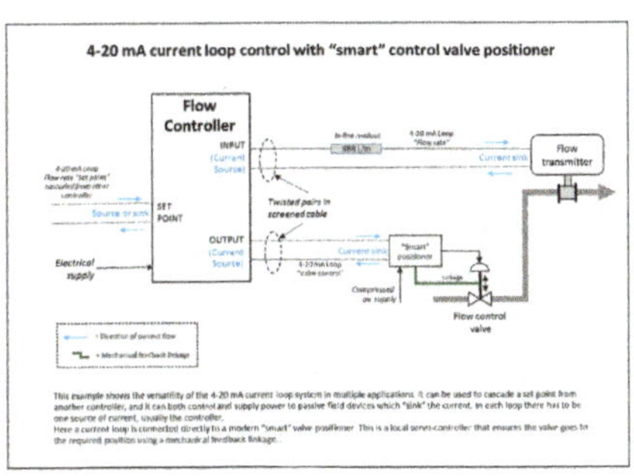

Example of current loops used for sensing and control transmission. Example of use of a modern "smart" valve positioner shown.

Other Applications

Electronic analog PID control loops were often found within more complex electronic systems, for example, the head positioning of a disk drive, the power conditioning of a power supply, or even the movement-detection circuit of a modern seismometer. Nowadays, electronic controllers have largely been replaced by digital controllers implemented with microcontrollers or FPGAs, which implement PID algorithms. However, analog PID controllers are still used in niche applications requiring high-bandwidth and low-noise performance, such as laser-diode controllers.

Present Day

Most modern PID controllers in industry are implemented in Distributed Control Systems (DCS), programmable logic controllers (PLCs) or as a panel-mounted digital controller. Software implementations have the advantages that they are relatively cheap and are flexible with respect to the implementation of the PID algorithm. PID temperature controllers are applied in industrial ovens, plastics injection machinery, hot-stamping machines and packing industry.

Control Loop Basics

A robotic arm can be moved and positioned by a control loop. By applying forward and reverse power to an electric motor to lift and lower the arm, it may be necessary to allow for the inertial mass of the arm, forces due to gravity, and to correct for external forces on the arm such as a load to lift or work to be done on an external object.

The sensed position is the process variable (PV). The desired position is called the setpoint (SP). The input to the process (the electric current in the motor) is the output from the PID controller. It is called either the manipulated variable (MV) or the control variable (CV). The difference between the present position and the setpoint is the error (e), which quantifies whether the arm is too low or too high and by how much.

By measuring the position (PV), and subtracting it from the setpoint (SP), the error (e) is found, and from it the controller calculates how much electric current to supply to the motor (MV). The obvious method is proportional control: the motor current is set in proportion to the existing error. A more complex control may include another term: derivative action. This considers the rate of change of error, supplying more or less electric current depending on how fast the error is approaching zero. Finally, integral action adds a third term, using the accumulated position error in the past to detect whether the position of the mechanical arm is settling out too low or too high and to set the electrical current in relation not only to the error but also the time for which it has persisted. An alternative formulation of integral action is to change the electric current in small persistent steps that are proportional to the current error. Over time the steps accumulate and add up dependent on past errors; this is the discrete-time equivalent to integration.

Applying too much impetus when the error is small and is reducing will lead to overshoot. After overshooting, if the controller were to apply a large correction in the opposite direction and repeatedly overshoot the desired position, the output would oscillate around the setpoint in either a constant, growing, or decaying sinusoid. If the amplitude of the oscillations increase with time, the system is unstable. If they decrease, the system is stable. If the oscillations remain at a constant magnitude, the system is marginally stable.

In the interest of achieving a controlled arrival at the desired position (SP) in a timely and accurate way, the controlled system needs to be critically damped. A well-tuned position control system will also apply the necessary currents to the controlled motor so that the arm pushes and pulls as necessary to resist external forces trying to move it away from the required position. The setpoint itself may be generated by an external system, such as a PLC or other computer system, so that it continuously varies depending on the work that the robotic arm is expected to do. A well-tuned PID control system will enable the arm to meet these changing requirements to the best of its capabilities.

If a controller starts from a stable state with zero error (PV = SP), then further changes by the controller will be in response to changes in other measured or unmeasured inputs to the process that affect the process, and hence the PV. Variables that affect the process other than the MV are known as disturbances. Generally controllers are used to reject disturbances and to implement setpoint changes. A change in load on the arm constitutes a disturbance to the robot arm control process.

In theory, a controller can be used to control any process which has a measurable output (PV), a known ideal value for that output (SP) and an input to the process (MV) that will affect the relevant PV. Controllers are used in industry to regulate temperature, pressure, force, feed rate, flow rate, chemical composition (component concentrations), weight, position, speed and practically every other variable for which a measurement exists.

PID Controller Theory

The PID control scheme is named after its three correcting terms, whose sum constitutes the manipulated variable (MV). The proportional, integral, and derivative terms are summed to calculate the output of the PID controller. Defining $u(t)$ as the controller output, the final form of the PID algorithm is

$$u(t) = \mathrm{MV}(t) = K_\mathrm{p} e(t) + K_\mathrm{i} \int_0^t e(\tau)\, d\tau + K_\mathrm{d} \frac{de(t)}{dt},$$

where

K_p is the proportional gain, a tuning parameter,

K_i is the integral gain, a tuning parameter,

K_d is the derivative gain, a tuning parameter,

$e(t) = \mathrm{SP} - \mathrm{PV}(t)$ is the error (SP is the setpoint, and PV(t) is the process variable),

t is the time or instantaneous time (the present),

τ is the variable of integration (takes on values from time 0 to the present t).

Equivalently, the transfer function in the Laplace domain of the PID controller is

$$L(s) = K_\mathrm{p} + K_\mathrm{i}/s + K_\mathrm{d} s,$$

where s is the complex frequency.

Proportional Term

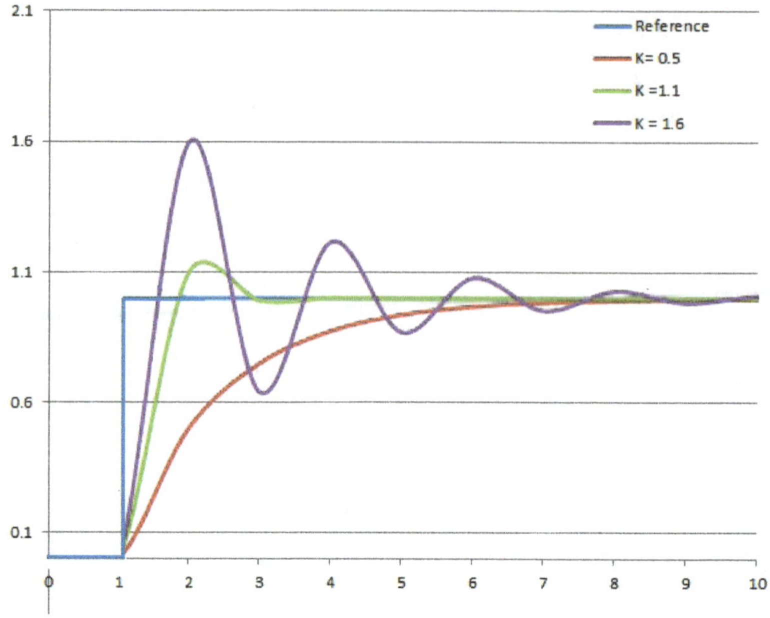

Plot of PV vs time, for three values of K_p (K_i and K_d held constant)

The proportional term produces an output value that is proportional to the current error value. The proportional response can be adjusted by multiplying the error by a constant K_p, called the proportional gain constant.

The proportional term is given by

$$P_{\mathrm{out}} = K_{\mathrm{p}} e(t).$$

A high proportional gain results in a large change in the output for a given change in the error. If the proportional gain is too high, the system can become unstable (see the section on loop tuning). In contrast, a small gain results in a small output response to a large input error, and a less responsive or less sensitive controller. If the proportional gain is too low, the control action may be too small when responding to system disturbances. Tuning theory and industrial practice indicate that the proportional term should contribute the bulk of the output change.

Steady-State Error

Because a non-zero error is required to drive it, a proportional controller generally operates with a so-called *steady-state error*.[a] Steady-state error (SSE) is proportional to the process gain and inversely proportional to proportional gain. SSE may be mitigated by adding a compensating bias term to the setpoint or output, or corrected dynamically by adding an integral term.

Integral Term

The contribution from the integral term is proportional to both the magnitude of the error and the duration of the error. The integral in a PID controller is the sum of the instantaneous error over time and gives the accumulated offset that should have been corrected previously. The accumulated error is then multiplied by the integral gain (K_i) and added to the controller output.

The integral term is given by

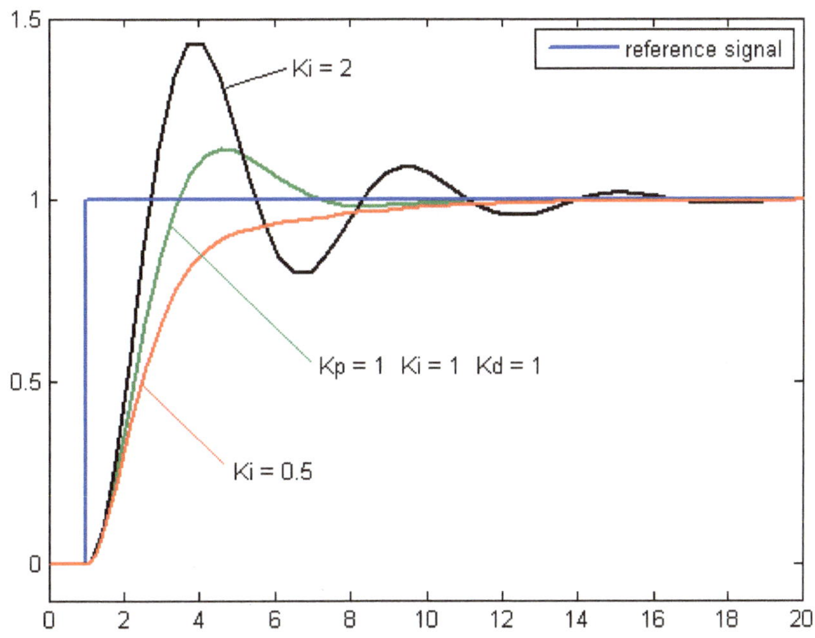

Plot of PV vs time, for three values of K_i (K_p and K_d held constant)

$$I_{\text{out}} = K_{\text{i}} \int_0^t e(\tau)\, d\tau.$$

The integral term accelerates the movement of the process towards setpoint and eliminates the residual steady-state error that occurs with a pure proportional controller. However, since the integral term responds to accumulated errors from the past, it can cause the present value to overshoot the setpoint value (see the section on loop tuning).

Derivative Term

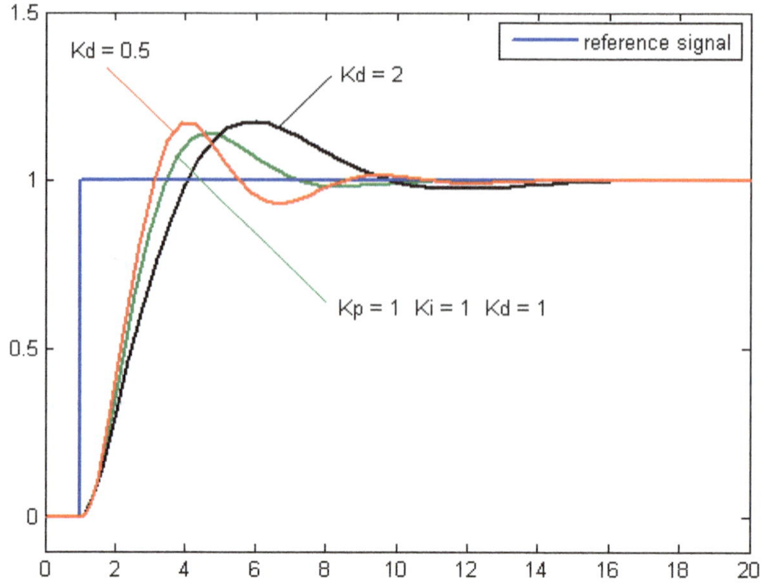

Plot of PV vs time, for three values of K_d (K_p and K_i held constant)

The derivative of the process error is calculated by determining the slope of the error over time and multiplying this rate of change by the derivative gain K_d. The magnitude of the contribution of the derivative term to the overall control action is termed the derivative gain, K_d.

The derivative term is given by

$$D_{\mathrm{out}} = K_{\mathrm{d}} \frac{de(t)}{dt}.$$

Derivative action predicts system behavior and thus improves settling time and stability of the system. An ideal derivative is not causal, so that implementations of PID controllers include an additional low-pass filtering for the derivative term to limit the high-frequency gain and noise. Derivative action is seldom used in practice though – by one estimate in only 25% of deployed controllers – because of its variable impact on system stability in real-world applications.

Loop Tuning

Tuning a control loop is the adjustment of its control parameters (proportional band/gain, integral gain/reset, derivative gain/rate) to the optimum values for the desired control response. Stability (no unbounded oscillation) is a basic requirement, but beyond that, different systems have different behavior, different applications have different requirements, and requirements may conflict with one another.

PID tuning is a difficult problem, even though there are only three parameters and in principle is simple to describe, because it must satisfy complex criteria within the limitations of PID control. There are accordingly various methods for loop tuning, and more sophisticated techniques are the subject of patents; this section describes some traditional manual methods for loop tuning.

Designing and tuning a PID controller appears to be conceptually intuitive, but can be hard in practice, if multiple (and often conflicting) objectives such as short transient and high stability are to be achieved. PID controllers often provide acceptable control using default tunings, but performance can generally be improved by careful tuning, and performance may be unacceptable with poor tuning. Usually, initial designs need to be adjusted repeatedly through computer simulations until the closed-loop system performs or compromises as desired.

Some processes have a degree of nonlinearity and so parameters that work well at full-load conditions don't work when the process is starting up from no-load; this can be corrected by gain scheduling (using different parameters in different operating regions).

Stability

If the PID controller parameters (the gains of the proportional, integral and derivative terms) are chosen incorrectly, the controlled process input can be unstable, i.e., its output diverges, with or without oscillation, and is limited only by saturation or mechanical breakage. Instability is caused by *excess* gain, particularly in the presence of significant lag.

Generally, stabilization of response is required and the process must not oscillate for any combination of process conditions and setpoints, though sometimes marginal stability (bounded oscillation) is acceptable or desired.

Mathematically, the origins of instability can be seen in the Laplace domain. The total loop transfer function is:

$$H(s) = \frac{K(s)G(s)}{1 + K(s)G(s)}$$

where

$K(s)$: PID transfer function

$G(s)$: Plant transfer function

The system is called unstable where the closed loop transfer function diverges for some s {\displaystyle s} s.[15] This happens for situations where K(s)G(s)=-1. Typically, this happens when K(s)G(s)|=1 with a 180 degree phase shift. Stability is guaranteed when K(s)G (s)<1 for frequencies that suffer high phase shifts. A more general formalism of this effect is known as the Nyquist stability criterion.

Optimum Behavior

The optimum behavior on a process change or setpoint change varies depending on the application.

Two basic requirements are *regulation* (disturbance rejection – staying at a given setpoint) and *command tracking* (implementing setpoint changes) – these refer to how well the controlled variable tracks the desired value. Specific criteria for command tracking include rise time and settling time. Some processes must not allow an overshoot of the process variable beyond the setpoint if, for example, this would be unsafe. Other processes must minimize the energy expended in reaching a new setpoint.

Overview of Methods

There are several methods for tuning a PID loop. The most effective methods generally involve the development of some form of process model, then choosing P, I, and D based on the dynamic model parameters. Manual tuning methods can be relatively time consuming, particularly for systems with long loop times.

The choice of method will depend largely on whether or not the loop can be taken "offline" for tuning, and on the response time of the system. If the system can be taken offline, the best tuning method often involves subjecting the system to a step change in input, measuring the output as a function of time, and using this response to determine the control parameters.

Choosing a tuning method

Method	Advantages	Disadvantages
Manual tuning	No math required; online.	Requires experienced personnel.
Ziegler–Nichols[b]	Proven method; online.	Process upset, some trial-and-error, very aggressive tuning.
Tyreus Luyben	Proven method; online.	Process upset, some trial-and-error, very aggressive tuning.
Software tools	Consistent tuning; online or offline - can employ computer-automated control system design (*CAutoD*) techniques; may include valve and sensor analysis; allows simulation before downloading; can support non-steady-state (NSS) tuning.	Some cost or training involved.
Cohen–Coon	Good process models.	Some math; offline; only good for first-order processes.
Åström-Hägglund	Can be used for auto tuning; amplitude is minimum so this method has lowest process upset	The process itself is inherently oscillatory.

PID Tuning Software

Most modern industrial facilities no longer tune loops using the manual calculation methods shown above. Instead, PID tuning and loop optimization software are used to ensure consistent results. These software packages will gather the data, develop process models, and suggest optimal tuning. Some software packages can even develop tuning by gathering data from reference changes.

Mathematical PID loop tuning induces an impulse in the system, and then uses the controlled system's frequency response to design the PID loop values. In loops with response times of several minutes, mathematical loop tuning is recommended, because trial and error can take days just to find a stable set of loop values. Optimal values are harder to find. Some digital loop controllers offer a self-tuning feature in which very small setpoint changes are sent to the process, allowing the controller itself to calculate optimal tuning values.

Other formulas are available to tune the loop according to different performance criteria. Many patented formulas are now embedded within PID tuning software and hardware modules.

Advances in automated PID Loop Tuning software also deliver algorithms for tuning PID Loops in a dynamic or Non-Steady State (NSS) scenario. The software will model the dynamics of a process, through a disturbance, and calculate PID control parameters in response.

Limitations of PID Control

While PID controllers are applicable to many control problems, and often perform satisfactorily without any improvements or only coarse tuning, they can perform poorly in some applications, and do not in general provide *optimal* control. The fundamental difficulty with PID control is that it is a feedback control system, with *constant* parameters, and no direct knowledge of the process, and thus overall performance is reactive and a compromise. While PID control is the best controller in an observer without a model of the process, better performance can be obtained by overtly modeling the actor of the process without resorting to an observer.

PID controllers, when used alone, can give poor performance when the PID loop gains must be reduced so that the control system does not overshoot, oscillate or hunt about the control setpoint value. They also have difficulties in the presence of non-linearities, may trade-off regulation versus response time, do not react to changing process behavior (say, the process changes after it has warmed up), and have lag in responding to large disturbances.

The most significant improvement is to incorporate feed-forward control with knowledge about the system, and using the PID only to control error. Alternatively, PIDs can be modified in more minor ways, such as by changing the parameters (either gain scheduling in different use cases or adaptively modifying them based on performance), improving measurement (higher sampling rate, precision, and accuracy, and low-pass filtering if necessary), or cascading multiple PID controllers.

Linearity

Another problem faced with PID controllers is that they are linear, and in particular symmetric. Thus, performance of PID controllers in non-linear systems (such as HVAC systems) is variable. For example, in temperature control, a common use case is active heating (via a heating element) but passive cooling (heating off, but no cooling), so overshoot can only be corrected slowly – it cannot be forced downward. In this case the PID should be tuned to be overdamped, to prevent or reduce overshoot, though this reduces performance (it increases settling time).

Noise in Derivative

A problem with the derivative term is that it amplifies higher frequency measurement or process noise that can cause large amounts of change in the output. It is often helpful to filter the measurements with a low-pass filter in order to remove higher-frequency noise components. As low-pass filtering and derivative control can cancel each other out, the amount of filtering is limited. So low noise instrumentation can be important. A nonlinear median filter may be used, which improves the filtering efficiency and practical performance. In some cases, the differential band can be turned off with little loss of control. This is equivalent to using the PID controller as a PI controller.

Modifications to the PID Algorithm

The basic PID algorithm presents some challenges in control applications that have been addressed by minor modifications to the PID form.

Integral Windup

One common problem resulting from the ideal PID implementations is integral windup. Following a large change in setpoint the integral term can accumulate an error larger than the maximal value for the regulation variable (windup), thus the system overshoots and continues to increase until this accumulated error is unwound. This problem can be addressed by:

- Disabling the integration until the PV has entered the controllable region
- Preventing the integral term from accumulating above or below pre-determined bounds
- Back-calculating the integral term to constrain the regulator output within feasible bounds.

Overshooting from Known Disturbances

For example, a PID loop is used to control the temperature of an electric resistance furnace where the system has stabilized. Now when the door is opened and something cold is put into the furnace the temperature drops below the setpoint. The integral function of the controller tends to compensate for error by introducing another error in the positive direction. This overshoot can be avoided by freezing of the integral function after the opening of the door for the time the control loop typically needs to reheat the furnace.

PI Controller

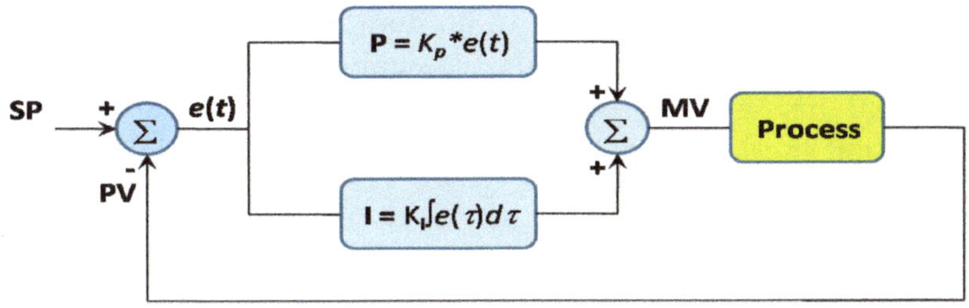

Basic block of a PI controller

A PI Controller (proportional-integral controller) is a special case of the PID controller in which the derivative (D) of the error is not used.

The controller output is given by

$$K_P\Delta + K_I \int \Delta \, dt$$

where Δ is the error or deviation of actual measured value (***PV***) from the setpoint (***SP***).

$\Delta = SP - PV$. A PI controller can be modelled easily in software such as Simulink or Xcos using a "flow chart" box involving Laplace operators:

$$C = \frac{G(1 + \tau s)}{\tau s}$$

where

$G = K_P$ = proportional gain

$G/\tau = K_I$ = integral gain

Setting a value for G is often a trade off between decreasing overshoot and increasing settling time.

The lack of derivative action may make the system more steady in the steady state in the case of noisy data. This is because derivative action is more sensitive to higher-frequency terms in the inputs.

Without derivative action, a PI-controlled system is less responsive to real (non-noise) and relatively fast alterations in state and so the system will be slower to reach setpoint and slower to respond to perturbations than a well-tuned PID system may be.

Deadband

Many PID loops control a mechanical device (for example, a valve). Mechanical maintenance can be a major cost and wear leads to control degradation in the form of either stiction or a deadband in the mechanical response to an input signal. The rate of mechanical wear is mainly a function of how often a device is activated to make a change. Where wear is a significant concern, the PID loop may have an output deadband to reduce the frequency of activation of the output (valve). This is accomplished by modifying the controller to hold its output steady if the change would be small (within the defined deadband range). The calculated output must leave the deadband before the actual output will change.

Setpoint Step Change

The proportional and derivative terms can produce excessive movement in the output when a system is subjected to an instantaneous step increase in the error, such as a large setpoint change. In the case of the derivative term, this is due to taking the derivative of the error, which is very large in the case of an instantaneous step change. As a result, some PID algorithms incorporate some of the following modifications:

Setpoint Ramping

In this modification, the setpoint is gradually moved from its old value to a newly specified value using a linear or first order differential ramp function. This avoids the discontinuity present in a simple step change.

Derivative of the Process Variable

In this case the PID controller measures the derivative of the measured process variable (PV), rather than the derivative of the error. This quantity is always continuous (i.e., never has a step change as a result of changed setpoint). This modification is a simple case of setpoint weighting.

Setpoint Weighting

Setpoint weighting adds adjustable factors (usually between 0 and 1) to the setpoint in the error in the proportional and derivative element of the controller. The error in the integral term must be the true control error to avoid steady-state control errors. These two extra parameters do not affect the response to load disturbances and measurement noise and can be tuned to improve the controller's setpoint response.

Feed-Forward

The control system performance can be improved by combining the feedback (or closed-loop) control of a PID controller with feed-forward (or open-loop) control. Knowledge about the system (such as the desired acceleration and inertia) can be fed forward and combined with the PID output to improve the overall system performance. The feed-forward value alone can often provide the major portion of the controller output. The PID controller primarily has to compensate whatever difference or *error* remains between the setpoint (SP) and the system response to the open loop control. Since the feed-forward output is not affected by the process feedback, it can never cause the control system to oscillate, thus improving the system response without affecting stability. Feed forward can be based on the setpoint and on extra measured disturbances. Setpoint weighting is a simple form of feed forward.

For example, in most motion control systems, in order to accelerate a mechanical load under control, more force is required from the actuator. If a velocity loop PID controller is being used to control the speed of the load and command the force being applied by the actuator, then it is beneficial to take the desired instantaneous acceleration, scale that value appropriately and add it to the output of the PID velocity loop controller. This means that whenever the load is being accelerated or decelerated, a proportional amount of force is commanded from the actuator regardless of the feedback value. The PID loop in this situation uses the feedback information to change the combined output to reduce the remaining difference between the process setpoint and the feedback value. Working together, the combined open-loop feed-forward controller and closed-loop PID controller can provide a more responsive control system.

Bumpless Operation

PID controllers are often implemented with a "bumpless" initialization feature that recalculates the integral accumulator term to maintain a consistent process output through parameter changes. A partial implementation is to store the integral of the integral gain times the error rather than storing the integral of the error and postmultiplying by the integral gain, which prevents discontinuous output when the I gain is changed, but not the P or D gains.

Other Improvements

In addition to feed-forward, PID controllers are often enhanced through methods such as PID gain scheduling (changing parameters in different operating conditions), fuzzy logic or computational verb logic. Further practical application issues can arise from instrumentation connected to the controller. A high enough sampling rate, measurement precision, and measurement accuracy are required to achieve adequate control performance. Another new method for improvement of PID controller is to increase the degree of freedom by using fractional order. The order of the integrator and differentiator add increased flexibility to the controller.

Cascade Control

One distinctive advantage of PID controllers is that two PID controllers can be used together to yield better dynamic performance. This is called cascaded PID control. In cascade control there are two PIDs arranged with one PID controlling the setpoint of another. A PID controller acts as outer loop controller, which controls the primary physical parameter, such as fluid level or velocity. The other controller acts as inner loop controller, which reads the output of outer loop controller as setpoint, usually controlling a more rapid changing parameter, flowrate or acceleration. It can be mathematically proventhat the working frequency of the controller is increased and the time constant of the object is reduced by using cascaded PID controllers.

For example, a temperature-controlled circulating bath has two PID controllers in cascade, each with its own thermocouple temperature sensor. The outer controller controls the temperature of the water using a thermocouple located far from the heater where it accurately reads the temperature of the bulk of the water. The error term of this PID controller is the difference between the desired bath temperature and measured temperature. Instead of controlling the heater directly, the outer PID controller sets a heater temperature goal for the inner PID controller. The inner PID controller controls the temperature of the heater using a thermocouple attached to the heater. The inner controller's error term is the difference between this heater temperature setpoint and the measured temperature of the heater. Its output controls the actual heater to stay near this setpoint.

The proportional, integral and differential terms of the two controllers will be very different. The outer PID controller has a long time constant – all the water in the tank needs to heat up or cool down. The inner loop responds much more quickly. Each controller can be tuned to match the physics of the system *it* controls – heat transfer and thermal mass of the whole tank or of just the heater – giving better total response.

REFERENCES

1. Zhong, Wan-Xie (2004). Duality System in Applied Mechanics and Optimal Control. Kluwer. p. 283. ISBN 978-1-4020-7880-4. The classical controller design methodology is iterative, and is effective for single-input, single-output linear time-invariant system analysis and design.
2. Ogata, Katsuhiko (2010). Modern Control Systems (Fifth ed.). Prentice Hall. p. 2. ISBN 978-0-13-615673-4. modern control theory, based on time-domain analysis and synthesis using state variables
3. Ang, K.H., Chong, G.C.Y., and Li, Y. (2005). PID control system analysis, design, and technology, IEEE Trans Control Systems Tech, 13(4), pp.559-576.
4. Phillips, C.l., Parr, J.M., & Riskin, E.A (2007). Signals, systems and Transforms. Prentice Hall. ISBN 0-13-041207-4.
5. Hespanha,J.P. (2009). Linear System Theory. Princeton university press. ISBN 0-691-14021-9.
6. Crutchfield, Steve (October 12, 2010), “The Joy of Convolution”, Johns Hopkins University, retrieved November 21, 2010
7. Vaidyanathan, P. P.; Chen, T. (May 1995). “Role of anticausal inverses in multirate filter banks — Part I: system theoretic fundamentals”. IEEE Trans. Signal Proc. 43 (6): 1090. Bibcode:1995ITSP...43.1090V. doi:10.1109/78.382395.
8. “Feedback and control systems” - JJ Di Steffano, AR Stubberud, IJ Williams. Schaums outline series, McGraw-Hill 1967
9. “Feedback and control systems” - JJ Di Steffano, AR Stubberud, IJ Williams.

Schaums outline series, McGraw-Hill 1967

10. Mayr, Otto (1970). The Origins of Feedback Control. Clinton, MA USA: The Colonial Press, Inc.
11. Mayr, Otto (1969). The Origins of Feedback Control. Clinton, MA USA: The Colonial Press, Inc.
12. Bennett, Stuart (1992). A history of control engineering, 1930-1955. IET. p. p. 48. ISBN 978-0-86341-299-8.
13. Antunes, Ricardo; Gonzalez, Vicente (March 3, 2015). "A Production Model for Construction: A Theoretical Framework". Buildings. 5 (1): 209–228. doi:10.3390/buildings5010209.
14. "Feedback and control systems" - JJ Di Steffano, AR Stubberud, IJ Williams. Schaums outline series, McGraw-Hill 1967
15. Mayr, Otto (1970). The Origins of Feedback Control. Clinton, MA USA: The Colonial Press, Inc.
16. Mayr, Otto (1969). The Origins of Feedback Control. Clinton, MA USA: The Colonial Press, Inc.
17. Bennett, Stuart (1992). A history of control engineering, 1930-1955. IET. p. p. 48. ISBN 978-0-86341-299-8.
18. Ang, K.H., Chong, G.C.Y., and Li, Y. (2005). PID control system analysis, design, and technology, IEEE Trans Control Systems Tech, 13(4), pp.559–576.
19. trim point
20. Donald M Wiberg. State space & linear systems. Schaum's outline series. McGraw Hill. ISBN 0-07-070096-6.
21. Terrell, William (1999). "Some fundamental control theory I: Controllability, observability, and duality —AND— Some fundamental control Theory II: Feedback linearization of single input nonlinear systems". Amer. Math. Monthly. 106: 705–719 and 812–828. doi:10.2307/2589614.
22. Gu Shi; et al. (2015). "Controllability of structural brain networks (Article Number 8414)". Nature Communications (6). doi:10.1038/ncomms9414. Lay summary. Here we use tools from control and network theories to offer a mechanistic explanation for how the brain moves between cognitive states drawn from the network organization of white matter microstructure.
23. Liu, Jie; Wilson Wang; Farid Golnaraghi; Eric Kubica (2010). "A novel fuzzy framework for nonlinear system control". Fuzzy Sets and Systems. 161 (21): 2746–2759. doi:10.1016/j.fss.2010.04.009.
24. Maxwell, J.C. (1868). "On Governors". Proceedings of the Royal Society of London. 16: 270–283. doi:10.1098/rspl.1867.0055. JSTOR 112510.
25. Routh, E.J.; Fuller, A.T. (1975). Stability of motion. Taylor & Francis.
26. Routh, E.J. (1877). A Treatise on the Stability of a Given State of Motion, Particularly Steady Motion: Particularly Steady Motion. Macmillan and co.
27. Hurwitz, A. (1964). "On The Conditions Under Which An Equation Has Only

Roots With Negative Real Parts". Selected Papers on Mathematical Trends in Control Theory.

28. Richard Bellman (1964) Control Theory, Scientific American 211(3):186–200

CHAPTER 7

Linear Control System

Chapter Contents

Access control systems are physical or electronic systems which are designed to control who has access to a network. The simplest example of a physical access control system is a door which can be locked, limiting people to one side of the door or the other. Electronic versions typically control network security, limiting which users are allowed to use resources on a computer system, for example.

INTRODUCTION

A control system is a device, or set of devices, that manages, commands, directs or regulates the behaviour of other devices or systems. They can range from a home heating controller using a thermostat controlling a boiler to large Industrial control systems which are used for controlling processes or machines.

In the most common form, the *feedback control system* it is desired to control a process, called the *plant*, so its output follows a *control signal*, which may be a fixed or changing value. The control system compares the output of the plant to the control signal, and applies the difference as an error signal to bring the output of the plant closer to the control signal.

A hydroelectric power station in Amerongen, Netherlands.

OVERVIEW

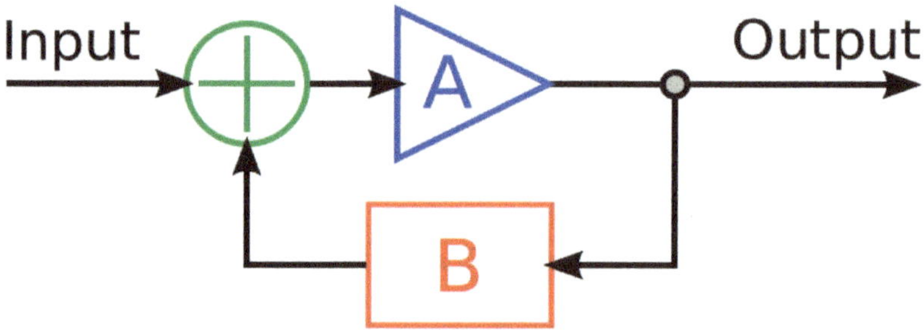

A basic feedback loop

The term "control system" may be applied to the essentially manual controls that allow an operator, for example, to close and open a hydraulic press, perhaps including logic so that it cannot be moved unless safety guards are in place.

An automatic sequential control system may trigger a series of mechanical actuators in the correct sequence to perform a task. For example, various electric and pneumatic transducers may fold and glue a cardboard box, fill it with product and then seal it in an automatic packaging machine. Programmable logic controllers are used in many cases such as this, but several alternative technologies exist.

In the case of linear feedback systems, a control loop, including sensors, control algorithms and actuators, is arranged in such a fashion as to try to regulate a variable at a setpoint or reference value. An example of this may increase the fuel supply to a furnace when a measured temperature drops. PID controllers are common and effective in cases such as this. Control systems that include some sensing of the results they are trying to achieve are making use of feedback and so can, to some extent, adapt to varying circumstances. Open-loop control systems do not make use of feedback, and run only in pre-arranged ways.

OPEN-LOOP AND CLOSED-LOOP CONTROL

There are two common classes of control systems, open loop control systems, and closed loop control systems.

In an open loop control system, the control action from the controller is independent of the "process output". A good example of this is a central heating boiler controlled only by a timer, so that heat is applied for a constant time, regardless of the temperature of the building. (The control action is the switching on/off of the boiler. The process output is the building temperature).

In a closed loop control system, the control action from the controller is dependent on the process output. In the case of the boiler analogy this would include a termperature thermostat to monitor the building temperature, and thereby feed back a signal to ensure the controller maintains the temperature set on the thermostat.

A closed loop controller therefore has a feedback loop which ensures the controller exerts a control action to give a process output the same as the "Reference input" or "set point". For this reason, closed loop controllers are also called feedback controllers.

Logic Control

An internal lift control panel.

Logic control systems for industrial and commercial machinery were historically implemented at mains voltage using interconnected relays, designed using ladder logic. Today, most such systems are constructed with programmable logic controllers (PLCs) or microcontrollers. The notation of ladder logic is still in use as a programming idiom for PLCs.

Logic controllers may respond to switches, light sensors, pressure switches, etc., and can cause the machinery to start and stop various operations. Logic systems are used to sequence mechanical operations in many applications. PLC software can be written in many different ways – ladder diagrams, SFC – sequential function charts or in language terms known as statement lists.

Examples include elevators, washing machines and other systems with interrelated stop-go operations.

Logic systems are quite easy to design, and can handle very complex operations. Some aspects of logic system design make use of Boolean logic.

ON–OFF CONTROL

A thermostat is a simple negative feedback controller: when the temperature (the "process variable" or PV) goes below a set point (SP), the heater is switched on. Another example could be a pressure switch on an air compressor. When the pressure (PV) drops below the threshold (SP), the pump is powered. Refrigerators and vacuum pumps contain similar mechanisms operating in reverse, but still providing negative feedback to correct errors.

Simple on–off feedback control systems like these are cheap and effective. In some cases, like the simple compressor example, they may represent a good design choice.

In most applications of on–off feedback control, some consideration needs to be given to other costs, such as wear and tear of control valves and perhaps other start-up costs when power is reapplied each time the PV drops. Therefore, practical on–off control systems are designed to include hysteresis which acts as a deadband, a region around the setpoint value in which no control action occurs. The width of deadband may be adjustable or programmable.

LINEAR CONTROL

Linear control systems use linear negative feedback to produce a control signal mathematically based on other variables, with a view to maintain the controlled process within an acceptable operating range.

The output from a linear control system into the controlled process may be in the form of a directly variable signal, such as a valve that may be 0 or 100% open or anywhere in between. Sometimes this is not feasible and so, after calculating the current required corrective signal, a linear control system may repeatedly switch an actuator, such as a pump, motor or heater, fully on and then fully off again, regulating the duty cycle using pulse-width modulation.

Proportional Control

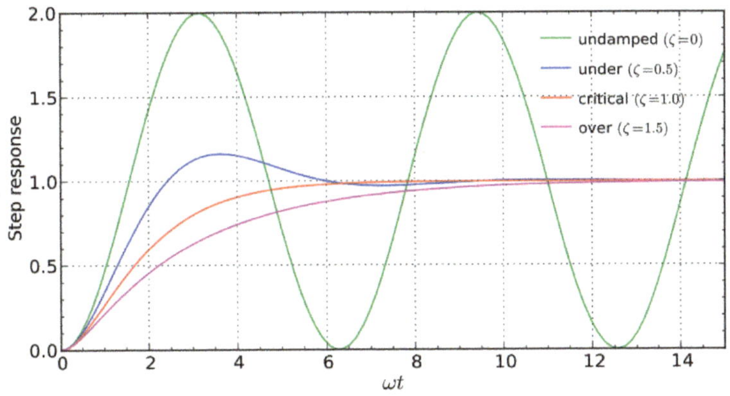

Step responses for a second order system defined by the transfer function $H(s) = \frac{\omega_n^2}{s^2 + 2\zeta\omega_n s + \omega_n^2}$, where where ζ is the damping ratio and ωn is the undamped natural frequency.is the undamped natural frequency.

When controlling the temperature of an industrialfurnace, it is usually better to control the opening of the fuel valve *in proportion to* the current needs of the furnace. This helps avoid thermal shocks and applies heat more effectively.

Proportional negative-feedback systems are based on the difference between the required set point (SP) and process value (PV). This difference is called the *error*. Power is applied in direct proportion to the current measured error, in the correct sense so as to tend to reduce the error and therefore avoid positive feedback. The amount of corrective action that is applied for a given error is set by the gain or sensitivity of the control system.

At low gains, only a small corrective action is applied when errors are detected. The system may be safe and stable, but may be sluggish in response to changing conditions. Errors will remain uncorrected for relatively long periods of time and the system is over-damped. If the proportional gain is increased, such systems become more responsive and errors are dealt with more quickly. There is an optimal value for the gain setting when the overall system is said to be critically damped. Increases in loop gain beyond this point lead to oscillations in the PV and such a system is under-damped.

In real systems, there are practical limits to the range of the manipulated variable (MV). For example, a heater can be off or fully on, or a valve can be closed or fully open. Adjustments to the gain simultaneously alter the range of error values over which the MV is between these limits. The width of this range, in units of the error variable and therefore of the PV, is called the *proportional band* (PB). While the gain is useful in mathematical treatments, the proportional band is often used in practical situations. They both refer to the same thing, but the PB has an inverse relationship to gain – higher gains result in narrower PBs, and *vice versa*.

Under-Damped Furnace Example

In the furnace example, suppose the temperature is increasing towards a set point at which, say, 50% of the available power will be required for steady-state. At low temperatures, 100% of available power is applied. When the process value (PV) is within, say 10° of the SP the heat input begins to be reduced by the proportional controller (note that this implies a 20° proportional band (PB) from full to no power input, evenly spread around the setpoint value). At the setpoint the controller will be applying 50% power as required, but stray stored heat within the heater sub-system and in the walls of the furnace will keep the measured temperature rising beyond what is required. At 10° above SP, we reach the top of the proportional band (PB) and no power is applied, but the temperature may continue to rise even further before beginning to fall back. Eventually as the PV falls back into the PB, heat is applied again, but now the heater and the furnace walls are too cool and the temperature falls too low before its fall is arrested, so that the oscillations continue.

Over-Damped Furnace Example

The temperature oscillations that an under-damped furnace control system produces are unacceptable for many reasons, including the waste of fuel and time (each oscillation cycle may take many minutes), as well as the likelihood of seriously overheating both the furnace and its contents.

Suppose that the gain of the control system is reduced drastically and it is restarted. As the temperature approaches, say 30° below SP (60° proportional band (PB)), the heat input begins to be reduced, the rate of heating of the furnace has time to slow and, as the heat is still further reduced, it eventually is brought up to set point, just as 50% power input is reached and the furnace is operating as required. There was some wasted time while the furnace crept to its final temperature using only 52% then 51% of available power, but at least no harm was done. By carefully increasing the gain (i.e. reducing the width of the PB) this over-damped and sluggish behavior can be improved until the system is critically damped for this SP temperature. Doing this is known as 'tuning' the control system. A well-tuned proportional furnace temperature control system will usually be more effective than on-off control, but will still respond more slowly than the furnace could under skillful manual control.

PID Control

Apart from sluggish performance to avoid oscillations, another problem with proportional-only control is that power application is always in direct proportion to the error. In the example above we assumed that the set temperature could be maintained with 50% power. What happens if the furnace is required in a different application where a higher set temperature will require 80% power to maintain it? If the gain was finally set to a 50° PB, then 80% power will not be applied unless the furnace is 15° below setpoint, so for this other application the operators will have to remember always to set the setpoint temperature 15° higher than actually needed. This 15° figure is not completely constant either: it will depend on the surrounding ambient temperature, as well as other factors that affect heat loss from or absorption within the furnace.

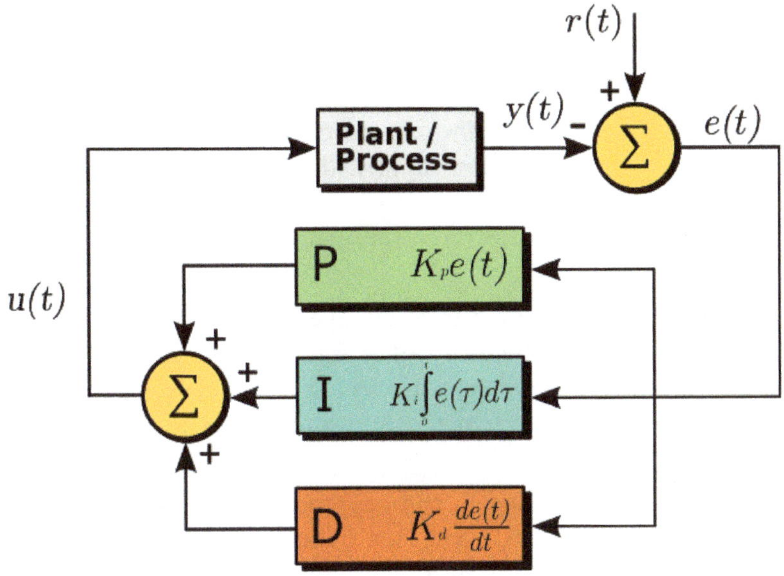

A block diagram of a PID controller

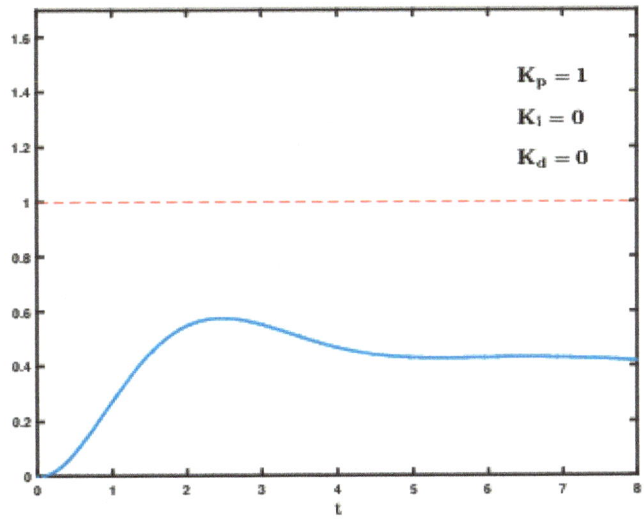

Effects of varying PID parameters (K_p,K_i,K_d) on the step response of a system.

To resolve these two problems, many feedback control schemes include mathematical extensions to improve performance. The most common extensions lead to proportional-integral-derivative control, or PID control.

Derivative Action

The derivative part is concerned with the rate-of-change of the error with time: If the measured variable approaches the setpoint rapidly, then the actuator is backed off early to allow it to coast to the required level; conversely if the measured value begins to move rapidly away from the setpoint, extra effort is applied—in proportion to that rapidity—to try to maintain it.

Derivative action makes a control system behave much more intelligently. On control systems like the tuning of the temperature of a furnace, or perhaps the motion-control of a heavy item like a gun or camera on a moving vehicle, the derivative action of a well-tuned PID controller can allow it to reach and maintain a setpoint better than most skilled human operators could.

If derivative action is over-applied, it can lead to oscillations too. An example would be a PV that increased rapidly towards SP, then halted early and seemed to "shy away" from the setpoint before rising towards it again.

Integral Action

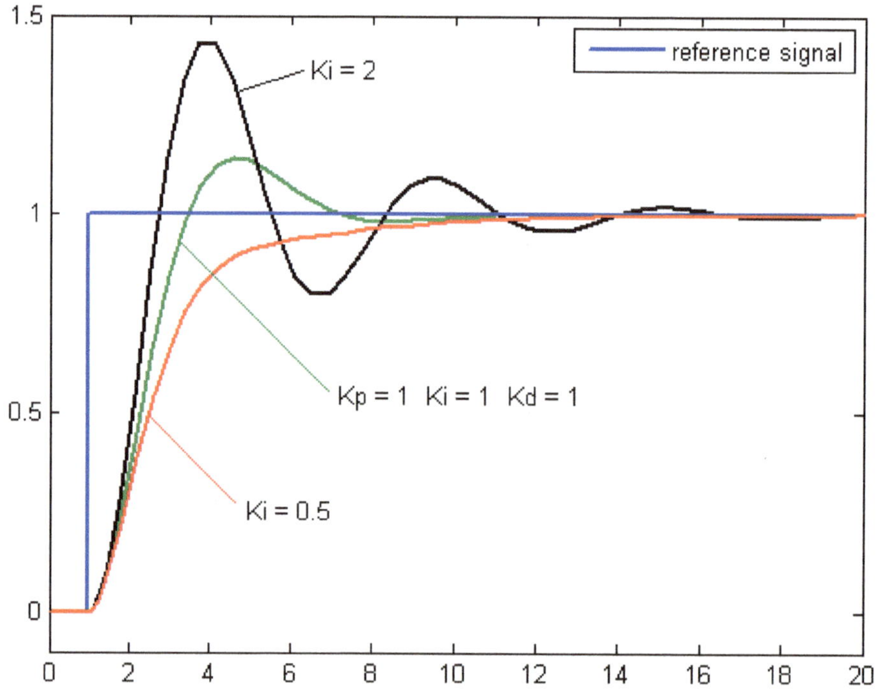

Change of response of second order system to a step input for varying Ki values.

The integral term magnifies the effect of long-term steady-state errors, applying ever-increasing effort until they reduce to zero. In the example of the furnace above working at various temperatures, if the heat being applied does not bring the furnace up to setpoint, for whatever reason, integral action increasingly *moves* the proportional band relative to the setpoint until the PV error is reduced to zero and the setpoint is achieved.

Ramp UP % Per Minute

Some controllers include the option to limit the "ramp up % per minute". This option can be very helpful in stabilizing small boilers (3 MBTUH), especially during the summer, during light loads. A utility boiler "unit may be required to change load at a rate of as much as 5% per minute (IEA Coal Online - 2, 2007)".

Other Techniques

It is possible to filter the PV or error signal. Doing so can reduce the response of the system to undesirable frequencies, to help reduce instability or oscillations. Some feedback systems will oscillate at just one frequency. By filtering out that frequency, more "stiff" feedback can be applied, making the system more responsive without shaking itself apart.

Feedback systems can be combined. In cascade control, one control loop applies control algorithms to a measured variable against a setpoint, but then provides a varying setpoint to another control loop rather than affecting process variables directly. If a system has several different measured variables to be controlled, separate control systems will be present for each of them.

Control engineering in many applications produces control systems that are more complex than PID control. Examples of such fields include fly-by-wire aircraft control systems, chemical plants, and oil refineries. Model predictive control systems are designed using specialized computer-aided-design software and empirical mathematical models of the system to be controlled.

Fuzzy Logic

Fuzzy logic is an attempt to apply the easy design of logic controllers to the control of complex continuously varying systems. Basically, a measurement in a fuzzy logic system can be partly true, that is if yes is 1 and no is 0, a fuzzy measurement can be between 0 and 1.

The rules of the system are written in natural language and translated into fuzzy logic. For example, the design for a furnace would start with: 'If the temperature is too high, reduce the fuel to the furnace. If the temperature is too low, increase the fuel to the furnace.'

Measurements from the real world (such as the temperature of a furnace) are converted to values between 0 and 1 by seeing where they fall on a triangle. Usually, the tip of the triangle is the maximum possible value which translates to 1. Fuzzy logic, then, modifies Boolean logic to be arithmetical. Usually the 'not' operation is 'output = 1 - input,' the 'and' operation is 'output = input.1 multiplied by input.2,' and 'or' is 'output

= 1 - ((1 - input.1) multiplied by (1 - input.2))'. This reduces to Boolean arithmetic if values are restricted to 0 and 1, instead of allowed to range in the unit interval [0,1].

The last step is to 'defuzzify' an output. Basically, the fuzzy calculations make a value between zero and one. That number is used to select a value on a line whose slope and height converts the fuzzy value to a real-world output number. The number then controls real machinery.

If the triangles are defined correctly and rules are right the result can be a good control system.

When a robust fuzzy design is reduced into a single, quick calculation, it begins to resemble a conventional feedback loop solution and it might appear that the fuzzy design was unnecessary. However, the fuzzy logic paradigm may provide scalability for large control systems where conventional methods become unwieldy or costly to derive.

Fuzzy electronics is an electronic technology that uses fuzzy logic instead of the two-value logic more commonly used in digital electronics.

PHYSICAL IMPLEMENTATIONS

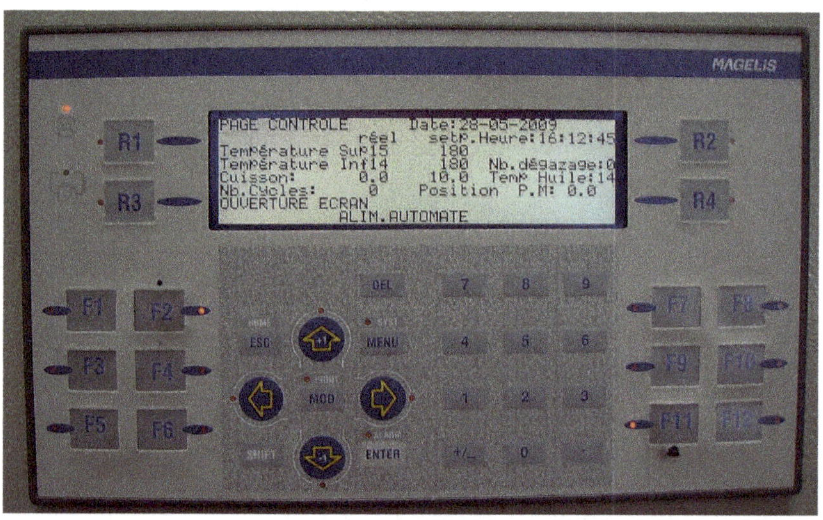

A control panel of a hydraulic heat press machine.

Since modern small microprocessors are so cheap (often less than $1 US), it's very common to implement control systems, including feedback loops, with computers, often in an embedded system. The feedback controls are simulated by having the computer make periodic measurements and then calculate from this stream of measurements (see digital signal processing, sampled data systems).

Computers emulate logic devices by making measurements of switch inputs, calculating a logic function from these measurements and then sending the results out to electronically controlled switches.

Logic systems and feedback controllers are usually implemented with programmable logic controllers which are devices available from electrical supply houses. They include a little computer and a simplified system for programming. Most often they are programmed with personal computers.

Logic controllers have also been constructed from relays, hydraulic and pneumatic devices as well as electronics using both transistors and vacuum tubes (feedback controllers can also be constructed in this manner).

REFERENCES

1. "Feedback and control systems" - JJ Di Steffano, AR Stubberud, IJ Williams. Schaums outline series, McGraw-Hill 1967
2. Kuphaldt, Tony R. "Chapter 6 LADDER LOGIC". Lessons In Electric Circuits -- Volume IV. Retrieved 22 September 2010.
3. Brady, Ian. "Programmable logic controllers - benefits and applications" (PDF). PLCs. Retrieved 5 December 2011.
4. http://www.seeei.org.il/prdFiles/2702_desc3.pdf ABB: Power Generation Energy Efficient Design of Auxiliary Systems in Fossil-Fuel Power Plants, Page 262, Section: Load Following

CHAPTER 8

Nonlinear Control

Chapter Contents

- **Introduction**
- **Properties Of Nonlinear Systems**
- **Analysis And Control Of Nonlinear Systems**

Nonlinear models naturally occur in most areas of engineering (mechanical, electrical, chemical) and are traditionally dealt with by linear approximations. However, by using the nonlinear model, better controllers can be designed that take into account the nonlinear behavior.

For affine models, you can automatically find a transformation that makes the system linear, allowing for the full suite of linear analysis and design functionality to be used. For general nonlinear models, automatic approximation schemes allow one to reduce to linear or affine systems or directly design a full information regulator.

INTRODUCTION

Nonlinear control theory is the area of control theory which deals with systems that are nonlinear, time-variant, or both. Control theory is an interdisciplinary branch of engineering and mathematics that is concerned with the behavior of dynamical systems with inputs, and how to modify the output by changes in the input using feedback. The system to be controlled is called the "plant". In order to make the output of a system follow a desired reference signal a controller is designed which compares the output of the plant to the desired output, and provides feedback to the plant to modify the output to bring it closer to the desired output. Control theory is divided into two branches:

Linear control theory applies to systems made of linear devices; which means they obey the superposition principle; the output of the device is proportional to its input. Systems with this property are governed by linear differential equations. A major subclass is systems which in addition have parameters which do not change with time, called *linear time invariant* (LTI) systems. These systems are amenable to powerful frequency domain mathematical techniques of great generality, such as the Laplace transform, Fourier transform, and Z transform, root-locus, Bode plot, and Nyquist stability criterion. These lead to a description of the system using terms like bandwidth, frequency response, eigenvalues, gain, resonant frequencies, poles, and zeros, which give solutions for system response and design techniques to most problems of interest.

Nonlinear control theory covers a wider class of systems that do not obey the superposition principle. It applies to more real-world systems, because all real control systems are nonlinear. These systems are often governed by nonlinear differential equations. The mathematical techniques which have been developed to handle them are more rigorous and much less general, often applying only to narrow categories of systems. These include limit cycle theory, Poincaré maps, Liapunov stability theory, and describing functions. If only solutions near a stable point are of interest, nonlinear systems can often be linearized by approximating them by a linear system obtained by expanding the nonlinear solution in a series, and then linear techniques can be used. Nonlinear systems are often analyzed using numerical methods on computers, for example by simulating their operation using a simulation language. Even if the plant is linear, a nonlinear controller can often have attractive features such as simpler implementation, faster speed, more accuracy, or reduced control energy, which justify the more difficult design procedure.

An example of a nonlinear control system is a thermostat-controlled heating system. A building heating system such as a furnace has a nonlinear response to changes in temperature; it is either "on" or "off", it does not have the fine control in response to temperature differences that a proportional (linear) device would have. Therefore, the furnace is off until the temperature falls below the "turn on" setpoint of the thermostat, when it turns on. Due to the heat added by the furnace, the temperature increases until it reaches the "turn off" setpoint of the thermostat, which turns the furnace off, and the cycle repeats. This cycling of the temperature about the desired temperature is called a *limit cycle*, and is characteristic of nonlinear control systems.

PROPERTIES OF NONLINEAR SYSTEMS

Some properties of nonlinear dynamic systems are

- They do not follow the principle of superposition (linearity and homogeneity).
- They may have multiple isolated equilibrium points.
- They may exhibit properties such as limit cycle, bifurcation, chaos.
- Finite escape time: Solutions of nonlinear systems may not exist for all times.

ANALYSIS AND CONTROL OF NONLINEAR SYSTEMS

There are several well-developed techniques for analyzing nonlinear feedback systems:

- Describing function method
- Phase plane method
- Lyapunov stability analysis
- Singular perturbation method
- Popov criterion (described in *The Lur'e Problem* below)
- Center manifold theorem
- Small-gain theorem
- Passivity analysis

Describing Function

In control systems theory, the describing function (DF) method, developed by Nikolay Mitrofanovich Krylov and Nikolay Bogoliubov in the 1930s, and extended by Ralph Kochenburger is an approximate procedure for analyzing certain nonlinear control problems. It is based on quasi-linearization, which is the approximation of the non-linear system under investigation by a linear time-invariant (LTI) transfer function that depends on the amplitude of the input waveform. By definition, a transfer function of a true LTI system cannot depend on the amplitude of the input function because an LTI system is linear. Thus, this dependence on amplitude generates a family of linear systems that are combined in an attempt to capture salient features of the non-linear system behavior. The describing function is one of the few widely applicable methods for designing nonlinear systems, and is very widely used as a standard mathematical tool for analyzing limit cycles in closed-loop controllers, such as industrial process controls, servomechanisms, and electronic oscillators.

The Method

Consider feedback around a discontinuous (but piecewise continuous) nonlinearity (e.g., an amplifier with saturation, or an element with deadband effects) cascaded with a slow stable linear system. Depending on the amplitude of the output of the linear system, the feedback presented to the nonlinearity will be in a different continuous region. As the output of the linear system decays, the nonlinearity may move into a different continuous region. This switching from one continuous region to another can generate periodic oscillations. The describing function method attempts to predict characteristics of those oscillations (e.g., their fundamental frequency) by assuming that the slow system acts like a low-pass or bandpass filter that concentrates all energy around a single frequency. Even if the output waveform has several modes, the method can still provide intuition about properties like frequency and possibly amplitude; in this case, the describing function method can be thought of as describing the sliding mode of the feedback system.

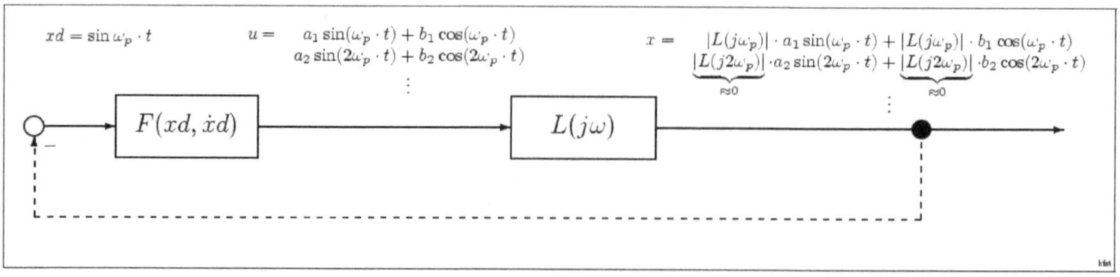

Nonlinear system in the state of harmonic balance

Using this low-pass assumption, the system response can be described by one of a family of sinusoidal waveforms; in this case the system would be characterized by a sine input describing function (SIDF) $H(A, j\omega)$ giving the system response to an input consisting of a sine wave of amplitude A and frequency w. This SIDF is a modification of the transfer function $H(j\omega)$ used to characterize linear systems. In a quasi-linear system, when the input is a sine wave, the output will be a sine wave of the same frequency but with a scaled amplitude and shifted phase as given by $H(A, j\omega)$. Many systems are approximately quasi-linear in the sense that although the response to a sine wave is not a pure sine wave, most of the energy in the output is indeed at the same frequency w as the input. This is because such systems may possess intrinsic low-pass or bandpass characteristics such that harmonics are naturally attenuated, or because external filters are added for this purpose. An important application of the SIDF technique is to estimate the oscillation amplitude in sinusoidal electronic oscillators.

Other types of describing functions that have been used are DFs for level inputs and for Gaussian noise inputs. Although not a complete description of the system, the DFs often suffice to answer specific questions about control and stability. DF methods are best for analyzing systems with relatively weak nonlinearities. In addition the higher order sinusoidal input describing functions (HOSIDF), describe the response of a class of nonlinear systems at harmonics of the input frequency of a sinusoidal input. The HOSIDFs are an extension of the SIDF for systems where the nonlinearities are significant in the response.

Caveats

Although the describing function method can produce reasonably accurate results for a wide class of systems, it can fail badly for others. For example, the method can fail if the system emphasizes higher harmonics of the nonlinearity. Such examples have been presented by Tzypkin for bang–bang systems. A fairly similar example is a closed-loop oscillator consisting of a non-inverting Schmitt trigger followed by an *inverting* integrator that feeds back its output to the Schmitt trigger's input. The output of the Schmitt trigger is going to be a square waveform, while that of the integrator (following it) is going to have a triangle waveform with peaks coinciding with the transitions in the square wave. Each of these two oscillator stages lags the signal exactly by 90 degrees (relative to its input). If one were to perform DF analysis on this circuit, the triangle wave at the Schmitt trigger's input would be replaced by its fundamental (sine wave), which passing through the trigger would cause a phase shift of less than 90 degrees (because the sine wave would trigger it sooner than the triangle wave does) so the system would appear not to oscillate in the same (simple) way.

Also, in the case where the conditions for Aizerman's or Kalman conjectures are fulfilled, there are no periodic solutions by describing function method, but counterexamples with periodic solutions (hidden oscillation) are well known. Therefore, the application of the describing function method requires additional justification.

Phase Plane

In applied mathematics, in particular the context of nonlinear system analysis, a phase plane is a visual display of certain characteristics of certain kinds of differential equations; a coordinate plane with axes being the values of the two state variables, say (x, y), or (q, p) etc. (any pair of variables). It is a two-dimensional case of the general n-dimensional phase space.

The phase plane method refers to graphically determining the existence of limit cycles in the solutions of the differential equation.

The solutions to the differential equation are a family of functions. Graphically, this can be plotted in the phase plane like a two-dimensional vector field. Vectors representing the derivatives of the points with respect to a parameter (say time t), that is $(dx/dt, dy/dt)$, at representative points are drawn. With enough of these arrows in place the system behaviour over the regions of plane in analysis can be visualized and limit cycles can be easily identified.

The entire field is the *phase portrait*, a particular path taken along a flow line (i.e. a path always tangent to the vectors) is a *phase path*. The flows in the vector field indicate the time-evolution of the system the differential equation describes.

In this way, phase planes are useful in visualizing the behaviour of physical systems; in particular, of oscillatory systems such as predator-prey models . In these models the phase paths can "spiral in" towards zero, "spiral out" towards infinity, or reach neutrally stable situations called centres where the path traced out can be either circular, elliptical, or ovoid, or some variant thereof. This is useful in determining if the dynamics are stable or not. Other examples of oscillatory systems are certain chemical reactions with multiple steps, some of which involve dynamic equilibria rather than reactions that go to completion. In such cases one can model the rise and fall of reactant and product concentration (or mass, or amount of substance) with the correct differential equations and a good understanding of chemical kinetics.

Example of a Linear System

A two-dimensional system of linear differential equations can be written in the form:

$$\frac{dx}{dt} = Ax + By$$
$$\frac{dy}{dt} = Cx + Dy$$

which can be organized into a matrix equation:

$$\frac{d}{dt}\begin{pmatrix} x \\ y \end{pmatrix} = \begin{pmatrix} A & B \\ C & D \end{pmatrix}\begin{pmatrix} x \\ y \end{pmatrix}$$
$$\frac{d\mathbf{x}}{dt} = \mathbf{A}\mathbf{x}.$$

where $\mathbf{A}$ is the 2×2 coefficient matrix above, and $\mathbf{x} = (x, y)$ is a coordinate vector of two independent variables.

Such systems may be solved analytically, for this case by integrating:

$$\frac{dy}{dx} = \frac{Cx + Dy}{Ax + By}$$

although the solutions are implicit functions in x and y, and are difficult to interpret.

Solving using Eigenvalues

More commonly they are solved with the coefficients of the right hand side written in matrix form using eigenvalues λ, given by the determinant:

$$\det(\mathbf{A} - \lambda\mathbf{I}) = 0$$

and eigenvectors:

$$\mathbf{A}\mathbf{x} = \lambda\mathbf{x}$$

The eigenvalues represent the powers of the exponential components and the eigenvectors are coefficients. If the solutions are written in algebraic form, they express the fundamental multiplicative factor of the exponential term. Due to the nonuniqueness of eigenvectors, every solution arrived at in this way has undetermined constants $c_1, c_2, \ldots c_n$.

The general solution is:

$$x = \begin{bmatrix} k_1 \\ k_2 \end{bmatrix} c_1 e^{\lambda_1 t} + \begin{bmatrix} k_3 \\ k_4 \end{bmatrix} c_2 e^{\lambda_2 t}.$$

where λ_1 and λ_2 are the eigenvalues, and (k_1, k_2), (k_3, k_4) are the basic eigenvectors. The constants c_1 and c_2 account for the nonuniqueness of eigenvectors and are not solvable unless an initial condition is given for the system.

The above determinant leads to the characteristic polynomial:

$$\lambda^2 - (A + D)\lambda + (AD - BC) = 0$$

which is just a quadratic equation of the form:

$$\lambda^2 - p\lambda + q = 0$$

where;

$$p = A + D = \mathrm{tr}(\mathbf{A}),$$

("tr" denotes trace) and

$$q = AD - BC = \det(\mathbf{A}).$$

The explicit solution of the eigenvalues are then given by the quadratic formula:

$$\lambda = \frac{1}{2}(p \pm \sqrt{\Delta})$$

where

$$\Delta = p^2 - 4q.$$

Eigenvectors and Nodes

The eigenvectors and nodes determine the profile of the phase paths, providing a pictorial interpretation of the solution to the dynamical system, as shown next.

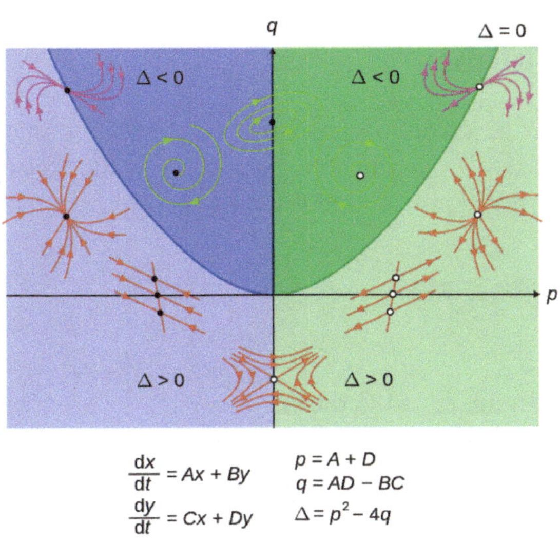

Classification of equilibrium points of a linearautonomous system. These profiles also arise for non-linear autonomous systems in linearized approximations.

The phase plane is then first set-up by drawing straight lines representing the two eigenvectors (which represent stable situations where the system either converges towards those lines or diverges away from them). Then the phase plane is plotted by using full lines instead of direction field dashes. The signs of the eigenvalues will tell how the system's phase plane behaves:

- If the signs are opposite, the intersection of the eigenvectors is a saddle point.
- If the signs are both positive, the eigenvectors represent stable situations that the system diverges away from, and the intersection is an unstable node.
- If the signs are both negative, the eigenvectors represent stable situations that the system converges towards, and the intersection is a stable node.

The above can be visualized by recalling the behavior of exponential terms in differential equation solutions.

Repeated Eigenvalues

This example covers only the case for real, separate eigenvalues. Real, repeated eigenvalues require solving the coefficient matrix with an unknown vector and the first eigenvector to generate the second solution of a two-by-two system. However, if the matrix is symmetric, it is possible to use the orthogonal eigenvector to generate the second solution.

Complex Eigenvalues

Complex eigenvalues and eigenvectors generate solutions in the form of sines and cosines as well as exponentials. One of the simplicities in this situation is that only one of the eigenvalues and one of the eigen vectors is needed to generate the full solution set for the system.

Singular Perturbation

In mathematics, a singular perturbation problem is a problem containing a small parameter that cannot be approximated by setting the parameter value to zero. More precisely, the solution cannot be uniformly approximated by an asymptotic expansion

$$\varphi(x) \approx \sum_{n=0}^{N} \delta_n(\varepsilon)\psi_n(x)$$

as $\varepsilon \to 0$. Here ε is the small parameter of the problem and $\delta_n(\varepsilon)$ are a sequence of functions of ε of increasing order, such as $\delta_n(\varepsilon) = \varepsilon^n$. This is in contrast to regular perturbation problems, for which a uniform approximation of this form can be obtained. Singularly perturbed problems are generally characterized by dynamics operating on multiple scales. Several classes of singular perturbations are outlined below.

Methods of Analysis

A perturbed problem whose solution can be approximated on the whole problem domain, whether space or time, by a single asymptotic expansion has a regular perturbation. Most often in applications, an acceptable approximation to a regularly perturbed problem is found by simply replacing the small parameter ε by zero everywhere in the problem statement. This corresponds to taking only the first term of the expansion, yielding an approximation that converges, perhaps slowly, to the true solution as ε decreases. The solution to a singularly perturbed problem cannot be approximated in this way: As seen in the examples below, a singular perturbation generally occurs when a problem's small parameter multiplies its highest operator. Thus naively taking the parameter to be zero changes the very nature of the problem. In the case of differential equations, boundary conditions cannot be satisfied; in algebraic equations, the possible number of solutions is decreased.

Singular perturbation theory is a rich and ongoing area of exploration for mathematicians, physicists, and other researchers. The methods used to tackle problems in this field are many. The more basic of these include the method of matched asymptotic expansions and WKB approximation for spatial problems, and in time, the Poincaré-Lindstedt method, the method of multiple scales and periodic averaging. For books on singular perturbation in ODE and PDE›s, see for example Holmes, *Introduction to Perturbation Methods*, Hinch, *Perturbation methods* or Bender and Orszag, *Advanced Mathematical Methods for Scientists and Engineers*.

Examples of Singular Perturbative Problems

Each of the examples described below shows how a naive perturbation analysis, which assumes that the problem is regular instead of singular, will fail. Some show how the problem may be solved by more sophisticated singular methods.

Vanishing Coefficients In Ordinary Differential Equations

Differential equations that contain a small parameter that premultiplies the highest order term typically exhibit boundary layers, so that the solution evolves in two different scales. For example, consider the boundary value problem

$$\varepsilon u''(x) + u'(x) = -e^{-x}, \quad 0 < x < 1$$
$$u(0) = 0, \quad u(1) = 1.$$

Its solution when $\varepsilon = 0.1$ is the solid curve shown below. Note that the solution changes rapidly near the origin. If we naively set $\varepsilon = 0$, we would get the solution labelled "outer" below which does not model the boundary layer, for which x is close to zero. For more details that show how to obtain the uniformly valid approximation, see method of matched asymptotic expansions.

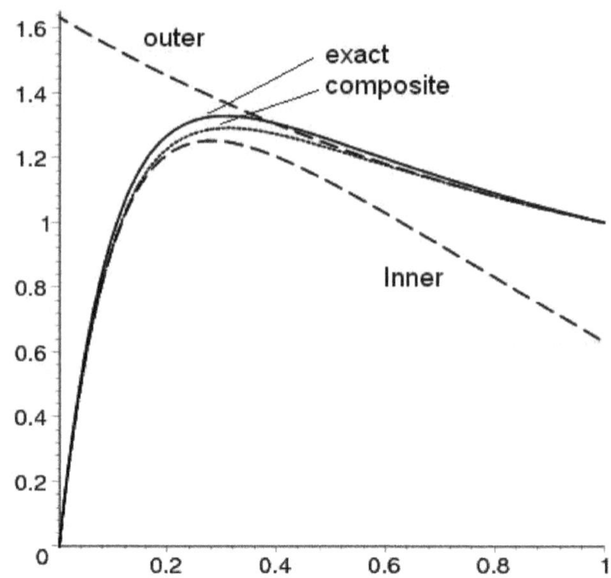

Examples in Time

An electrically driven robot manipulator can have slower mechanical dynamics and faster electrical dynamics, thus exhibiting two time scales. In such cases, we can divide the system into two subsystems, one corresponding to faster dynamics and other corresponding to slower dynamics, and then design controllers for each one of them separately. Through a singular perturbation technique, we can make these two subsystems independent of each other, thereby simplifying the control problem.

Consider a class of system described by following set of equations:

$$\begin{aligned} \dot{x}_1 &= f_1(x_1, x_2) + \varepsilon g_1(x_1, x_2, \varepsilon), \\ \varepsilon \dot{x}_2 &= f_2(x_1, x_2) + \varepsilon g_2(x_1, x_2, \varepsilon), \\ x_1(0) &= a_1, x_2(0) = a_2, \end{aligned}$$

with $0 < \varepsilon \ll 1$. The second equation indicates that the dynamics of x_2 is much faster than that of x_1. A theorem due to Tikhonov states that, with the correct conditions on the system, it will initially and very quickly approximate the solution to the equations

$$\begin{aligned} \dot{x}_1 &= f_1(x_1, x_2), \\ f_2(x_1, x_2) &= 0, \\ x_1(0) &= a_1, \end{aligned}$$

on some interval of time and that, as ε decreases toward zero, the system will approach the solution more closely in that same interval.

Examples in Space

In fluid mechanics, the properties of a slightly viscous fluid are dramatically different outside and inside a narrow boundary layer. Thus the fluid exhibits multiple spatial scales.

Reaction-diffusion systems in which one reagent diffuses much more slowly than another can form spatial patterns marked by areas where a reagent exists, and areas where it does not, with sharp transitions between them. In ecology, predator-prey models such as

$$u_t = \varepsilon u_{xx} + uf(u) - vg(u),$$
$$v_t = v_{xx} + vh(u),$$

where u is the prey and v is the predator, have been shown to exhibit such patterns.

Center Manifold

In mathematics, the **center manifold** of an equilibrium point of a dynamical system consists of orbits whose behavior around the equilibrium point is not controlled by either the attraction of the stable manifold or the repulsion of the unstable manifold. The first step when studying equilibrium points of dynamical systems is to linearize the system. The eigenvectors corresponding to eigenvalues with negative real part form the stable eigenspace, which gives rise to the stable manifold. Similarly, eigenvalues with positive real part yield the unstable manifold.

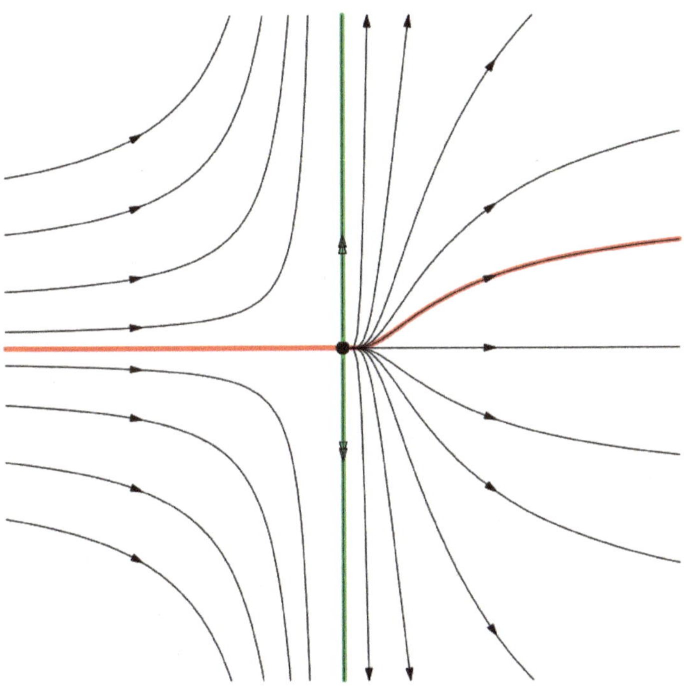

Center (red) and unstable (green) invariant manifolds of saddle-node equilibrium point

This concludes the story if the equilibrium point is hyperbolic (i.e., all eigenvalues of the linearization have nonzero real part). However, if there are eigenvalues whose real part is zero, then these give rise to the center manifold. If the eigenvalues are precisely zero, rather than just real part being zero, then these more specifically give rise to a slow manifold. The behavior on the center (slow) manifold is generally not determined by the linearization and thus is more difficult to study. Center manifolds play an important role in: bifurcation theory because interesting behavior takes place on the center manifold; and multiscale mathematics because the long time dynamics often are attracted to a relatively simple center manifold.

Definition

Let $\frac{d\mathbf{x}}{dt} = \mathbf{f}(\mathbf{x})$ be a dynamical system with equilibrium point x*. The linearization of the system near the equilibrium point is

$$\frac{d\mathbf{x}}{dt} = A\mathbf{x}, \quad \text{where Jacobian } A = \frac{d\mathbf{f}}{d\mathbf{x}}(\mathbf{x}^*).$$

The matrix A defines three main subspaces:

- the stable subspace, which is spanned by the generalized eigenvectors corresponding to the eigenvalues λ with $\mathrm{Re}\,\lambda < 0$;
- the unstable subspace, which is spanned by the generalized eigenvectors corresponding to the eigenvalues λ with $\mathrm{Re}\,\lambda < 0$;
- the center subspace, which is spanned by the generalized eigenvectors corresponding to the eigenvalues λ with $\mathrm{Re}\,\lambda < 0$;

Depending upon the application, other subspaces of interest include center-stable, center-unstable, sub-center, slow, and fast subspaces. These subspaces are all invariant subspaces of the linearized equation.

Corresponding to the linearized system, the nonlinear system has invariant manifolds, each consisting of sets of orbits of the nonlinear system.

- An invariant manifold tangent to the stable subspace and with the same dimension is the stable manifold.
- The unstable manifold is of the same dimension and tangent to the unstable subspace.
- A center manifold is of the same dimension and tangent to the center subspace. If, as is common, the eigenvalues of the center subspace are all precisely zero, rather than just real part zero, then a center manifold is often called a slow manifold.

Center Manifold Theorems

The center manifold existence theorem states that if the right-hand side function $\mathbf{f}(\mathbf{x})$ is C^r times continuously differentiable), then at every equilibrium point there exists a neighborhood of some finite size in which there is at least one of

a unique C^r **stable** manifold,

a unique C^r **unstable** manifold,

and a (not necessarily unique) C^{r-1} **center** manifold.

In example applications, a nonlinear coordinate transform to a normal form can clearly separate these three manifolds. A web service currently undertakes the necessary computer algebra for a range of finite-dimensional systems. In the case when the unstable manifold does not exist, center manifolds are often relevant to modelling. The center manifold emergence theorem then says that the neighborhood may be chosen so that all solutions of the system staying in the neighborhood tend exponentially quickly to some solution y(t) on the center manifold. That is, $\mathbf{x}(t) = \mathbf{y}(t) + O(e^{-\beta' t}) \quad \text{as } t \to \infty$, for some rate β’. This theorem asserts that for a wide variety of initial conditions the solutions of the full system decay exponentially quickly to a solution on the relatively low dimensional center manifold.

A third theorem, the approximation theorem, asserts that if an approximate expression for such invariant manifolds, say $\mathbf{x} = \mathbf{X}(\mathbf{s})$, satisfies the differential equation for the system to residuals $O(|\mathbf{s}|^p)$ as $\mathbf{s} \to \mathbf{0}$, then the invariant manifold is approximated by $\mathbf{x} = \mathbf{X}(\mathbf{s})$ to an error of the same order, namely $O(|\mathbf{s}|^p)$.

However, some applications, such as to shear dispersion, require an infinite-dimensional center manifold. The most general and powerful theory was developed by Aulbach and Wanner They addressed non-autonomous dynamical systems $\frac{d\mathbf{x}}{dt} = \mathbf{f}(\mathbf{x}, t)$ in infinite dimensions, with potentially infinite dimensional stable, unstable and center manifolds. Further, they usefully generalized the definition of the manifolds so that the center manifold is associated with eigenvalues such that $|\mathrm{Re}\,\lambda| \leq \alpha$, the stable manifold with eigenvalues $\mathrm{Re}\,\lambda \leq -\beta < -r\alpha$, and unstable manifold with eigenvalues $\mathrm{Re}\,\lambda \geq \beta > r\alpha$. . They proved existence of these manifolds, and the emergence of a center manifold, via nonlinear coordinate transforms. Potzsche and Rasmussen established a corresponding approximation theorem for such infinite dimensional, non-autonomous systems.

Center Manifold and the Analysis of Nonlinear Systems

As the stability of the equilibrium correlates with the “stability” of its manifolds, the existence of a center manifold brings up the question about the dynamics on the center manifold. This is analyzed by the center manifold reduction, which, in combination with some system parameter μ, leads to the concepts of bifurcations.

Correspondingly, two web services currently undertake the necessary computer algebra to construct just the center manifold for a wide range of finite-dimensional systems (provided they are in multinomial form).

- One web service constructs slow manifolds for systems which are linearly diagonalised, but which may be non-autonomous or stochastic.

- Another web service constructs center manifolds for systems with general linearisation, but only for autonomous systems.

Examples

The Wikipedia entry on slow manifolds gives more examples.

A Simple Example

Consider the system

$$\dot{x} = x^2, \quad \dot{y} = y.$$

The unstable manifold at the origin is the y axis, and the stable manifold is the trivial set $\{(0, 0)\}$. Any orbit not on the stable manifold satisfies an equation on the form $y = Ae^{-1/x}$ for some real constant A. It follows that for any real A, we can create a center manifold by piecing together the curve $y = Ae^{-1/x}$ for $x > 0$ with the negative x axis (including the origin). Moreover, all center manifolds have this potential non-uniqueness, although often the non-uniqueness only occurs in unphysical complex values of the variables.

Delay Differential Equations often have Hopf Bifurcations

Another example shows how a center manifold models the Hopf bifurcation that occurs for parameter $a \approx 4$ in the delay differential equation $dx/dt = -ax(t-1) - 2x^2 - x^3$. . Strictly, the delay makes this DE infinite-dimensional.

Fortunately, we may approximate such delays by the following trick that keeps the dimensionality finite. Define $u_1(t) = x(t)$ and approximate the time delayed variable, $x(t-1) \approx u_3(t)$, by using the intermediaries $du_2/dt = 2(u_1 - u_2)$ and $du_3/dt = 2(u_2 - u_3)$. .

For parameter near critical,a = 4 + a, the delay differential equation is then approximated by the system

$$\frac{d\mathbf{u}}{dt} = \begin{bmatrix} 0 & 0 & -4 \\ 2 & -2 & 0 \\ 0 & 2 & -2 \end{bmatrix} \mathbf{u} + \begin{bmatrix} -\alpha u_3 - 2u_1^2 - u_1^3 \\ 0 \\ 0 \end{bmatrix}.$$

Copying and pasting the appropriate entries, the web service finds that in terms of a complex amplitude s(t) and its complex conjugate $\bar{s}(t)$, the center manifold

$$\mathbf{u} = \begin{bmatrix} e^{i2t}s + e^{-i2t}\bar{s} \\ \frac{1-i}{2}e^{i2t}s + \frac{1+i}{2}e^{-i2t}\bar{s} \\ -\frac{i}{2}e^{i2t}s + \frac{i}{2}e^{-i2t}\bar{s} \end{bmatrix} + O(\alpha + |s|^2)$$

and the evolution on the center manifold is

$$\frac{ds}{dt} = \left[\frac{1+2i}{10}\alpha s - \frac{3+16i}{15}|s|^2 s\right] + O(\alpha^2 + |s|^4)$$

This evolution shows the origin is linearly unstable for $\alpha > 0\ (a > 4)$, but the cubic nonlinearity then stabilises nearby limit cycles as in classic Hopf bifurcation.

Small-Gain Theorem

In nonlinear systems, the formalism of input-output stability is an important tool in studying the stability of interconnected systems since the gain of a system directly relates to how the norm of a signal increases or decreases as it passes through the system. The small-gain theorem gives a sufficient condition for finite-gain L stability of the feedback connection. The small gain theorem was proved by George Zames in 1966. It can be seen as a generalization of the Nyquist criterion to non-linear time varying MIMO systems (systems with multiple inputs and multiple outputs).

Theorem. Assume two stable systems S_1 and S_2 are connected in a feedback loop, then the closed loop system is input-output stable if $\|S_1\| \cdot \|S_2\| < 1$. (The norm can be the infinity norm, that is the size of the largest singular value of the transfer function over all frequencies. Also any induced Norm will lead to the same results).

Control design techniques for nonlinear systems also exist. These can be subdivided into techniques which attempt to treat the system as a linear system in a limited range of operation and use (well-known) linear design techniques for each region:

- Gain scheduling

In control theory, gain scheduling is an approach to control of non-linear systems that uses a family of linear controllers, each of which provides satisfactory control for a different operating point of the system.

One or more observable variables, called the *scheduling variables*, are used to determine what operating region the system is currently in and to enable the appropriate linear controller. For example, in an aircraft flight control system, the altitude and Mach number might be the scheduling variables, with different linear controller parameters available (and automatically plugged into the controller) for various combinations of these two variables.

A relatively large scope state of the art about gain scheduling has been published in (Survey of Gain-Scheduling Analysis & Design, D.J.Leith, WE.Leithead).

Those that attempt to introduce auxiliary nonlinear feedback in such a way that the system can be treated as linear for purposes of control design:

- Feedback linearization

Feedback linearization is a common approach used in controlling nonlinear systems. The approach involves coming up with a transformation of the nonlinear system into an equivalent linear system through a change of variables and a suitable control input. Feedback linearization may be applied to nonlinear systems of the form

$$\dot{x} = f(x) + g(x)u$$
$$y = h(x)$$

where $x \in \mathbb{R}^n$ is the state vector, $u \in \mathbb{R}^p$ is the vector of inputs, and $y \in \mathbb{R}^m$ is the vector of outputs. The goal is to develop a control input

$$u = a(x) + b(x)v$$

that renders a linear input–output map between the new input v and the output. An outer-loop control strategy for the resulting linear control system can then be applied.

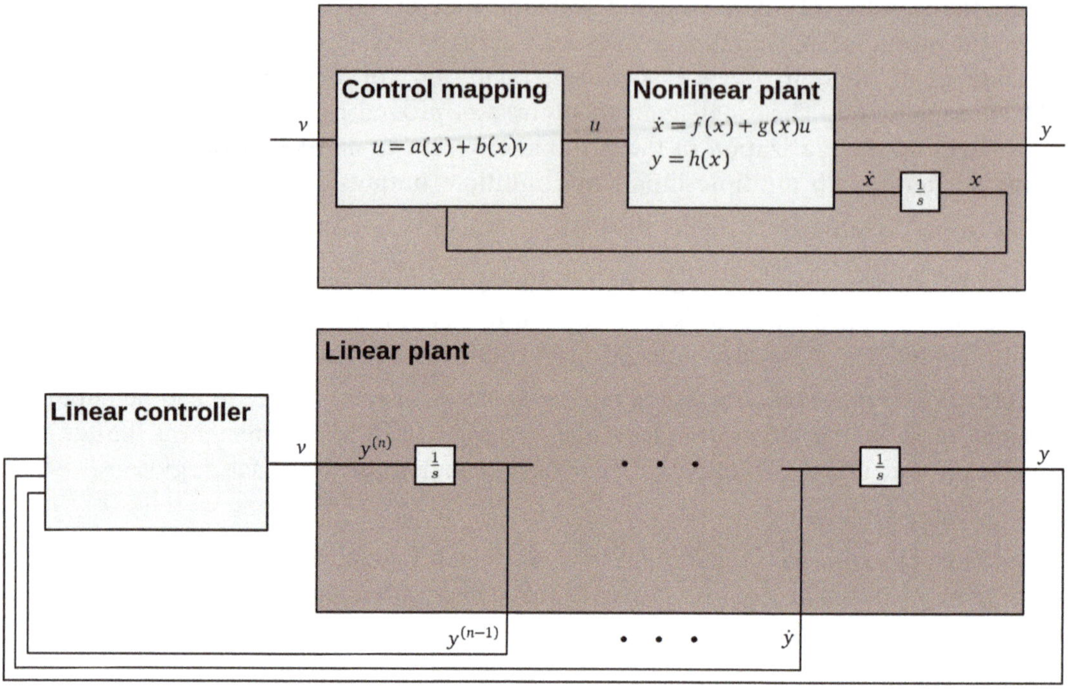

Block diagram illustrating the feedback linearization of a nonlinear system

Feedback Linearization of SISO Systems

Here, we consider the case of feedback linearization of a single-input single-output (SISO) system. Similar results can be extended to multiple-input multiple-output (MIMO) systems. In this case, $u \in \mathbb{R}$ and $y \in \mathbb{R}$. We wish to find a coordinate transformation $z = T(x)$ that transforms our system (1) into the so-called normal form which will reveal a feedback law of the form

$$u = a(x) + b(x)v$$

that will render a linear input–output map from the new input $v \in \mathbb{R}$ to the output y. To ensure that the transformed system is an equivalent representation of the original system, the transformation must be a diffeomorphism. That is, the transformation must not only be invertible (i.e., bijective), but both the transformation and its inverse must be smooth so that differentiability in the original coordinate system is preserved in the new coordinate system. In practice, the transformation can be only locally diffeomorphic, but the linearization results only hold in this smaller region.

We require several tools before we can solve this problem.

Lie Derivative

The goal of feedback linearization is to produce a transformed system whose states are the output y and its first n-1 derivatives. To understand the structure of this target system, we use the Lie derivative. Consider the time derivative of (2), which we can compute using the chain rule,

$$\begin{aligned}\dot{y} = \frac{\mathrm{d}\, h(x)}{\mathrm{d}\, t} &= \frac{\mathrm{d}\, h(x)}{\mathrm{d}\, x}\dot{x} \\ &= \frac{\mathrm{d}\, h(x)}{\mathrm{d}\, x}f(x) + \frac{\mathrm{d}\, h(x)}{\mathrm{d}\, x}g(x)u\end{aligned}$$

Now we can define the Lie derivative of $h(x)$ along $f(x)$ as,

$$L_f h(x) = \frac{\mathrm{d}\, h(x)}{\mathrm{d}\, x}f(x),$$

and similarly, the Lie derivative of $h(x)$ along $g(x)$ as,

$$L_g h(x) = \frac{\mathrm{d}\, h(x)}{\mathrm{d}\, x}g(x).$$

With this new notation, we may express $\dot{y}$ as,

$$\dot{y} = L_f h(x) + L_g h(x)u$$

Note that the notation of Lie derivatives is convenient when we take multiple derivatives with respect to either the same vector field, or a different one. For example,

$$L_f^2 h(x) = L_f L_f h(x) = \frac{\mathrm{d}(L_f h(x))}{\mathrm{d}\, x}f(x),$$

and

$$L_g L_f h(x) = \frac{\mathrm{d}(L_f h(x))}{\mathrm{d}\, x}g(x).$$

Relative Degree

In our feedback linearized system made up of a state vector of the outputy and its first n-1 derivatives, we must understand how the input u enters the system. To do this, we introduce the notion of relative degree. Our system given by (1) and (2) is said to have relative degree $r \in \mathbb{W}$ at a point x_0 if,

$$L_g L_f^k h(x) = 0 \qquad \forall x \text{ in a}$$
$$L_g L_f^{r-1} h(x_0) \neq 0$$

Considering this definition of relative degree in light of the expression of the time derivative of the output y, we can consider the relative degree of our system (1) and (2) to be the number of times we have to differentiate the output y before the input u appears explicitly. In an LTI system, the relative degree is the difference between the degree of the transfer function's denominator polynomial (i.e., number of poles) and the degree of its numerator polynomial (i.e., number of zeros).

Linearization by Feedback

For the discussion that follows, we will assume that the relative degree of the system is n. In this case, after differentiating the output n times we have,

$$\begin{aligned}
y &= h(x) \\
\dot{y} &= L_f h(x) \\
\ddot{y} &= L_f^2 h(x) \\
&\vdots \\
y^{(n-1)} &= L_f^{n-1} h(x) \\
y^{(n)} &= L_f^n h(x) + L_g L_f^{n-1} h(x) u
\end{aligned}$$

where the notation $y^{(n)}$ indicates the nth derivative of y. Because we assumed the relative degree of the system is n, the Lie derivatives of the form $L_g L_f^i h(x)$ for $i = 1, \ldots, n-2$ are all zero. That is, the input u has no direct contribution to any of the first n-1th derivatives.

The coordinate transformation T(x) that puts the system into normal form comes from the first n-1 derivatives. In particular,

$$z = T(x) = \begin{bmatrix} z_1(x) \\ z_2(x) \\ \vdots \\ z_n(x) \end{bmatrix} = \begin{bmatrix} y \\ \dot{y} \\ \vdots \\ y^{(n-1)} \end{bmatrix} = \begin{bmatrix} h(x) \\ L_f h(x) \\ \vdots \\ L_f^{n-1} h(x) \end{bmatrix}$$

transforms trajectories from the original x coordinate system into the new z coordinate system. So long as this transformation is a diffeomorphism, smooth trajectories in the original coordinate system will have unique counterparts in the z coordinate system that are also smooth. Those z trajectories will be described by the new system,

$$\begin{cases} \dot{z}_1 & = L_f h(x) = z_2(x) \\ \dot{z}_2 & = L_f^2 h(x) = z_3(x) \\ & \vdots \\ \dot{z}_n & = L_f^n h(x) + L_g L_f^{n-1} h(x) u \end{cases}.$$

Hence, the feedback control law

$$u = \frac{1}{L_g L_f^{n-1} h(x)} \left(-L_f^n h(x) + v\right)$$

renders a linear input–output map from v to $z_1 = y$. The resulting linearized system

$$\begin{cases} \dot{z}_1 & = z_2 \\ \dot{z}_2 & = z_3 \\ & \vdots \\ \dot{z}_n & = v \end{cases}$$

is a cascade of n integrators, and an outer-loop control v may be chosen using standard linear system methodology. In particular, a state-feedback control law of

$$v = -Kz \quad ,$$

where the state vector z is the output y and its first n- derivatives, results in the LTI system

$$\dot{z} = Az$$

with,

$$A = \begin{bmatrix} 0 & 1 & 0 & \dots & 0 \\ 0 & 0 & 1 & \dots & 0 \\ \vdots & \vdots & \vdots & \ddots & \vdots \\ 0 & 0 & 0 & \dots & 1 \\ -k_1 & -k_2 & -k_3 & \dots & -k_n \end{bmatrix}.$$

So, with the appropriate choice of k, we can arbitrarily place the closed-loop poles of the linearized system.

Unstable Zero Dynamics

Feedback linearization can be accomplished with systems that have relative degree less than n. However, the normal form of the system will include zero dynamics (i.e., states that are not observable from the output of the system) that may be unstable. In practice, unstable dynamics may have deleterious effects on the system (e.g., it may be dangerous for internal states of the system to grow unbounded). These unobservable states may be controllable or at least stable, and so measures can be taken to ensure these states do not cause problems in practice. Minimum phase systems provide some insight on zero dynamics.

And Lyapunov based methods:

- Lyapunov redesign
- Nonlinear damping
- Backstepping
- Sliding mode control

Nonlinear feedback analysis – The Lur'e problem

An early nonlinear feedback system analysis problem was formulated by A. I. Lur'e. Control systems described by the Lur'e problem have a forward path that is linear and time-invariant, and a feedback path that contains a memory-less, possibly time-varying, static nonlinearity.

The linear part can be characterized by four matrices (A,B,C,D), while the nonlinear part is Φ(y) with $\frac{\Phi(y)}{y} \in [a, b], \quad a < b \quad \forall y$ (a sector nonlinearity).

Absolute Stability Problem

Consider:

- (A,B) is controllable and (C,A) is observable
- two real numbers a, b with a<b, defining a sector for function Φ

The problem is to derive conditions involving only the transfer matrix H(s) and {a,b} such that x=0 is a globally uniformly asymptotically stable equilibrium of the system. This is known as the Lur›e problem. There are two well-known wrong conjections on absolute stability:

- The Aizerman›s conjecture
- The Kalman›s conjecture.

Aizerman's Conjecture

In nonlinear control, Aizerman's conjecture or Aizerman problem states that a linear system in feedback with a sector nonlinearity would be stable if the linear system is stable for any linear gain of the sector. This conjecture was proven false but led to the (valid) sufficient criteria on absolute stability.

Mathematical statement of Aizerman›s conjecture (Aizerman problem)

Consider a system with one scalar nonlinearity

$$\frac{dx}{dt} = Px + qf(e), \quad e = r^*x \quad x \in R^n,$$

*where P is a constant n×n-matrix, q, r are constant n-dimensional vectors, * is an operation of transposition, f(e) is scalar function, and f(0)=0. Suppose that the nonlinearity f is sector bounded, meaning that for some real* k_1 and k_2 with $k_1 < k_2$, the function f satisfies

$$k_1 < \frac{f(e)}{e} < k_2, \quad \forall\, e \neq 0.$$

Then Aizerman's conjecture is that the system is stable in large (i.e. unique stationary point is global attractor) if all linear systems with f(e)=ke, k ∈(k1,k2) are asymptotically stable.

There are counterexamples to Aizerman's conjecture such that nonlinearity belongs to the sector of linear stability and unique stable equilibrium coexists with a stable periodic solution—hidden oscillation

Strengthening of Aizerman's conjecture is Kalman's conjecture (or Kalman problem) where in place of condition on the nonlinearity it is required that the derivative of nonlinearity belongs to linear stability sector.

Kalman's Conjecture

Kalman's conjecture or Kalman problem is a disproved conjecture on absolute stability of nonlinear control system with one scalar nonlinearity, which belongs to the sector of linear stability. Kalman›s conjecture is a strengthening of Aizerman›s conjecture and is a special case of Markus–Yamabe conjecture. This conjecture was proven false but led to the (valid) sufficient criteria on absolute stability.

Mathematical Statement of Kalman's Conjecture (Kalman problem)

In 1957 R. E. Kalman in his paper stated the following:

If *f*(*e*) in Fig. 1 is replaced by constants *K* corresponding to all possible values of *f'*(*e*), and it is found that the closed-loop system is stable for all such *K*, then it intuitively clear that the system must be monostable; i.e., all transient solutions will converge to a unique, stable critical point.

Kalman's statement can be reformulated in the following conjecture:

Consider a system with one scalar nonlinearity

$$\frac{dx}{dt} = Px + qf(e), \quad e = r^*x \quad x \in R^n,$$

where P is a constant $n \times n$ matrix, q, r are constant n-dimensional vectors, $*$ is an operation of transposition, $f(e)$ is scalar function, and $f(0) = 0$. Suppose, $f(e)$ is a differentiable function and the following condition

$$k_1 < f'(e) < k_2.$$

is valid. Then Kalman's conjecture is that the system is stable in the large (i.e. a unique stationary point is a global attractor) if all linear systems with $f(e) = ke$, $k \in (k_1, k_2)$ are asymptotically stable.

In Aizerman's conjecture in place of the condition on the derivative of nonlinearity it is required that the nonlinearity itself belongs to the linear sector.

Kalman's conjecture is true for $n \leq 3$ and for $n > 3$ there are effective methods for construction of counterexamples: the nonlinearity derivative belongs to the sector of linear stability, and a unique stable equilibrium coexists with a stable periodic solution (hidden oscillation).

In discrete-time, the Kalman conjecture is only true for n=1, counterexamples for $n \geq 2$ can be constructed.

There are counterexamples to Aizerman›s and Kalman›s conjectures such that nonlinearity belongs to the sector of linear stability and unique stable equilibrium coexists with a stable periodic solution -- hidden oscillation.

There are two main theorems concerning the problem:

- The Circle criterion
- The Popov criterion.

which give sufficient conditions of absolute stability.

Popov Criterion

The sub-class of Lur'e systems studied by Popov is described by:

$$\begin{aligned} \dot{x} &= Ax + bu \\ \dot{\xi} &= u \\ y &= cx + d\xi \quad (1) \end{aligned}$$

$$u = -\phi(y) \quad (2)$$

where x ∈ R^n, ξ,u,y are scalars and A,b,c,d have commensurate dimensions. The nonlinear element Φ: R → R is a time-invariant nonlinearity belonging to *open sector* (0, ∞). This means that

$\Phi(0) = 0,\ y\ \Phi(y) > 0,\ \forall\ y \neq 0;$

The transfer function from u to y is given by

$$H(s) = \frac{d}{s} + c(sI - A)^{-1}b$$

Theorem: Consider the system (1)-(2) and suppose

1. A is Hurwitz
2. (A,b) is controllable
3. (A,c) is observable
4. d>0 and
5. $\Phi \in (0,\infty)$

then the system is globally asymptotically stable if there exists a number r>0 such that $\inf_{\omega \in R} \mathrm{Re}[(1+j\omega r)h(j\omega)] > 0$.

Things to be noted:

- The Popov criterion is applicable only to autonomous systems
- The system studied by Popov has a pole at the origin and there is no direct pass-through from input to output
- The nonlinearity Φ must satisfy an open sector condition

Theoretical Results in Nonlinear Control

Frobenius Theorem

The Frobenius theorem is a deep result in Differential Geometry. When applied to Nonlinear Control, it says the following: Given a system of the form

$$\dot{x} = \sum_{i=1}^{k} f_i(x)u_i(t)$$

where $x \in R^n$, $f_1, \ldots, f_k$ are vector fields belonging to a distribution Δ and $u_i(t)$ are control functions, the integral curves of x are restricted to a manifold of dimension m if $\text{span}(\Delta) = m$ and Δ is an involutive distribution.

REFERENCES

1. A.I. Lur'e and V. N. Postnikov, "On the theory of stability of control systems," Applied mathematics and mechanics, 8(3), 1944, (in Russian).
2. M. Vidyasagar, Nonlinear Systems Analysis, 2nd edition, Prentice Hall, Englewood Cliffs, New Jersey 07632.
3. A. Isidori, Nonlinear Control Systems, 3rd edition, Springer Verlag, London, 1995.
4. H. K. Khalil, Nonlinear Systems, 3rd edition, Prentice Hall, Upper Saddle River, New Jersey, 2002. ISBN 0-13-067389-7
5. B. Brogliato, R. Lozano, B. Maschke, O. Egeland, "Dissipative Systems Analysis and Control", Springer Verlag, London, 2nd edition, 2007.
6. Leonov G.A.; Kuznetsov N.V. (2011). "Algorithms for Searching for Hidden Oscillations in the Aizerman and Kalman Problems" (PDF). Doklady Mathematics. 84 (1): 475–481. doi:10.1134/S1064562411040120.
7. Bragin V.O.; Vagaitsev V.I.; Kuznetsov N.V.; Leonov G.A. (2011). "Algorithms for Finding Hidden Oscillations in Nonlinear Systems. The Aizerman and Kalman Conjectures and Chua's Circuits" (PDF). Journal of Computer and Systems Sciences International. 50 (5): 511–543. doi:10.1134/S106423071104006X.
8. Leonov G.A., Kuznetsov N.V. (2011). Sergio, Bittanti, ed. "Analytical-numerical methods for investigation of hidden oscillations in nonlinear control systems" (PDF). IFAC Proceedings Volumes (IFAC-PapersOnline). Proceedings of the 18th IFAC World Congress. 18 (1): 2494–2505. doi:10.3182/20110828-6-IT-1002.03315. ISBN 9783902661937.
9. Krylov, N. M.; N. Bogoliubov (1943). Introduction to Nonlinear Mechanics. Princeton, US: Princeton Univ. Press. ISBN 0691079854.
10. Jump up ^ Blaquiere, Austin. Nonlinear System Analysis. Elsevier Science. p. 177. ISBN 0323151663.
11. Jump up ^ Kochenburger, Ralph J. (January 1950). "A Frequency Response Method for Analyzing and Synthesizing Contactor Servomechanisms". Trans. of the AIEE. American Institute of Electrical Engineers. 69 (1): 270–284. doi:10.1109/t-aiee.1950.5060149. Retrieved June 18, 2013.
12. Jump up ^ Tsypkin, Yakov Z. (1984). Relay Control Systems. Cambridge: Univ Press.
13. Jump up ^ Boris Lurie; Paul Enright (2000). Classical Feedback Control: With MATLAB. CRC Press. pp. 298–299. ISBN 978-0-8247-0370-7.
14. Jump up ^ Leonov G.A.; Kuznetsov N.V. (2011). "Algorithms for Searching for Hidden Oscillations in the Aizerman and Kalman Problems" (PDF). Doklady Mathematics. 84 (1): 475–481. doi:10.1134/S1064562411040120.,
15. Jump up ^ "Aizerman's and Kalman's conjectures and describing function method" (PDF).
16. Jump up ^ Bragin V.O.; Vagaitsev V.I.; Kuznetsov N.V.; Leonov G.A. (2011). "Algorithms for Finding Hidden Oscillations in Nonlinear Systems. The

Aizerman and Kalman Conjectures and Chua's Circuits" (PDF). Journal of Computer and Systems Sciences International. 50 (4): 511–543. doi:10.1134/S106423071104006X.

17. Jump up ^ Leonov G.A.; Kuznetsov N.V. (2013). "Hidden attractors in dynamical systems. From hidden oscillations in Hilbert-Kolmogorov, Aizerman, and Kalman problems to hidden chaotic attractor in Chua circuits". International Journal of Bifurcation and Chaos. 23 (1): art. no. 1330002. doi:10.1142/S0218127413300024.

Distributed Control System

Chapter Contents

Distributed Control System is a specially designed control system used to control complex, large and geographically distributed applications in industrial processes. In this, controllers are distributed throughout the entire plant area.

These distributed controllers are connected to both field devices and operating PCs through high speed communication networks as shown in figure.

Discrete Field devices such as sensors and actuators are directly connected to input and output controller modules through communication bus. These field devices or smart instruments are capable of communicating with PLC's or other controllers while interacting with real world parameters like temperature, pressure, etc.

INTRODUCTION

A Distributed Control System (DCS) is a computerized control system for a process or plant, in which autonomous controllers are distributed throughout the system, but there is central operator supervisory control. This is in contrast to non-distributed control systems that use centralized controllers; either discrete controllers located at a central control room or within a central computer. The DCS concept increases reliability and reduces installation costs by localizing control functions near the process plant, but enables monitoring and supervisory control of the process remotely.

Distributed Control Systems first emerged in large, high value, safety critical process industries, and were attractive because the DCS manufacturer would supply both the local control level and central supervisory equipment as an integrated package, thus reducing design integration risk. Today the functionality of SCADA and **DCS** systems are very similar, but DCS tends to be used on large continuous process plants where high reliability and security is important, and the control room is not geographically remote.

STRUCTURE

The key attribute of a DCS is its reliability due to the distribution of the control processing around nodes in the system. This mitigates a single processor failure. If a processor fails, it will only affect one section of the plant process, as opposed to a failure of a central computer which would affect the whole process. This distribution of computing power local to the field Input/Output (I/O) field connection racks also ensures fast controller processing times by removing possible network and central processing delays.

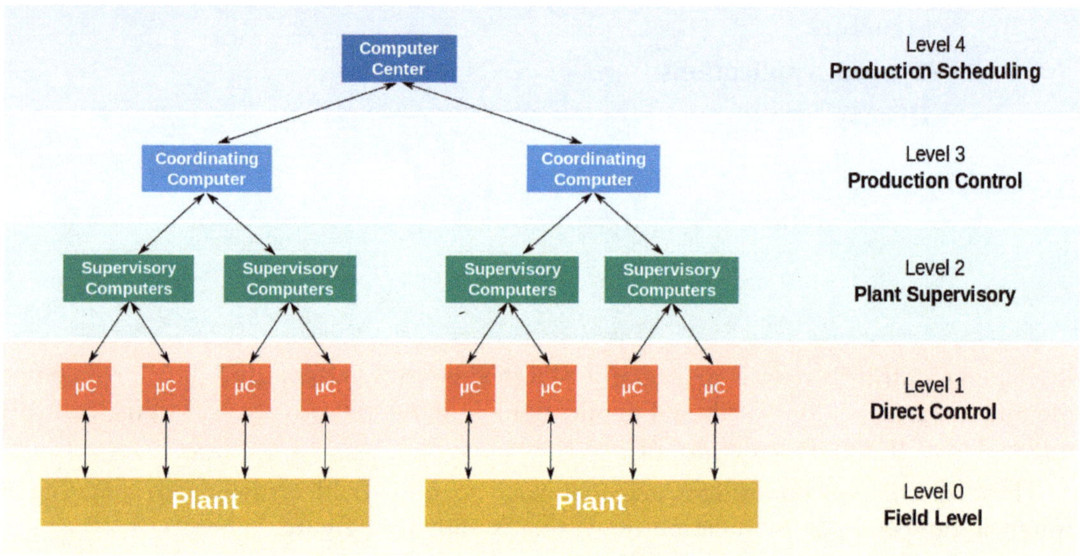

Functional levels of a manufacturing control operation.

The accompanying diagram is a general model which shows functional manufacturing levels using computerised control.

Referring to the diagram;

- Level 0 contains the field devices such as flow and temperature sensors, and final control elements, such as control valves
- Level 1 contains the industrialized Input/Output (I/O) modules, and their associated distributed electronic processors.
- Level 2 contains the supervisory computers, which collect information from processor nodes on the system, and provide the operator control screens.
- Level 3 is the production control level, which does not directly control the process, but is concerned with monitoring production and monitoring targets
- Level 4 is the production scheduling level.

Levels 1 and 2 are the functional levels of a traditional DCS, in which all equipment are part of an integrated system from a single manufacturer.

Levels 3 and 4 are not strictly process control in the traditional sense, but where production control and scheduling takes place.

Technical Points

The processor nodes and operator graphical displays are connected over proprietary or industry standard networks, and network reliability is increased by dual redundancy cabling over diverse routes. This distributed topology also reduces the amount of field cabling by siting the I/O modules and their associated processors close to the process plant.

4-20 mA current loop control with "smart" control valve positioner

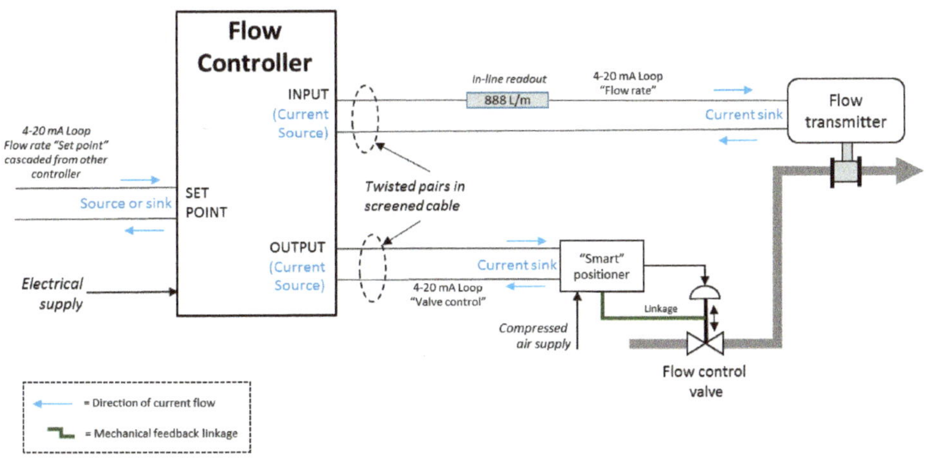

This example shows the versatility of the 4-20 mA current loop system in multiple applications. It can be used to cascade a set point from another controller, and it can both control and supply power to passive field devices which "sink" the current. In each loop there has to be one source of current, usually the controller.

Here a current loop is connected directly to a modern "smart" valve positioner. This is a local servo-controller that ensures the valve goes to the required position using a mechanical feedback linkage..

Example of a continuous flow control loop. Signaling is by industry standard 4-20 mA current loops, and a "smart" valve positioner ensures the control valve operates correctly.

The processors receive information from input modules, process the information and decide control actions to be signaled by the output modules. The field inputs and outputs can be analog signals e.g. 4~ 20mA dc current loop or 2 state signals that switch either "on" or "off", such as relay contacts or a semiconductor switch.

DCSs are connected to sensors and actuators and use setpoint control to control the flow of material through the plant. A typical application is a PID controller fed by a flow meter and using a control valve as the final control element. The DCS sends the setpoint required by the process to the controller which instructs a valve to operate so that the process reaches and stays at the desired setpoint. (see 4-20 mA schematic for example).

Large oil refineries and chemical plants have several thousand I/O points and employ very large DCS. Processes are not limited to fluidic flow through pipes, however, and can also include things like paper machines and their associated quality controls, variable speed drives and motor control centers, cement kilns, mining operations, ore processing facilities, and many others.

DCSs in very high reliability applications can have dual redundant processors with "hot" switch over on fault, to enhance the reliability of the control system.

Although 4-20 mA has been the main field signaling standard, modern DCS systems can also support fieldbus digital protocols, such as Foundation Fieldbus, profibus, HART, Modbus, PC Link etc, and other digital communication protocols such as modbus.

Modern DCSs also support neural networks and Fuzzy logic applications. Recent research focuses on the synthesis of optimal distributed controllers, which optimizes a certain H-infinity or the H 2 control criterion.

TYPICAL APPLICATIONS

Distributed Control Systems (DCS) are dedicated systems used in manufacturing processes that are continuous or batch-oriented.

Processes where a DCS might be used include:

- Chemical plants
- Petrochemical (oil) and refineries
- Pulp and Paper Mills
- Boiler controls and power plant systems
- Nuclear power plants
- Environmental control systems
- Water management systems
- Water treatment plants
- Sewage treatment plants
- Food and food processing
- Agro chemical and fertilizer
- Metal and mines
- Automobile manufacturing
- Metallurgical process plants
- Pharmaceutical manufacturing
- Sugar refining plants

HISTORY

A pre-DCS era central control room. Whilst the controls are centralized in one place, they are still discrete and not integrated into one system.

A DCS control room where plant information and controls are displayed on computer graphics screens. The operators are seated as they can view and control any part of the process from their screens, whilst retaining a plant overview.

Evolution of Process Control Operations

Process control of large industrial plants has evolved through many stages. Initially, control would be from panels local to the process plant. However this required a large manpower resource to attend to these dispersed panels, and there was no overall view of the process. The next logical development was the transmission of all plant measurements to a permanently-manned central control room. Effectively this was the centralization of all the localized panels, with the advantages of lower manning levels and easier overview of the process. Often the controllers were behind the control room panels, and all automatic and manual control outputs were transmitted back to plant. However, whilst providing a central control focus, this arrangement was inflexible as each control loop had its own controller hardware, and continual operator movement within the control room was required to view different parts of the process.

With the coming of electronic processors and graphic displays it became possible to replace these discrete controllers with computer-based algorithms, hosted on a network of input/output racks with their own control processors. These could be distributed around plant, and communicate with the graphic display in the control room or rooms. The Distributed Control System was born.

The introduction of DCSs allowed easy interconnection and re-configuration of plant controls such as cascaded loops and interlocks, and easy interfacing with other production computer systems. It enabled sophisticated alarm handling, introduced automatic event logging, removed the need for physical records such as chart recorders, allowed the control racks to be networked and thereby located locally to plant to reduce cabling runs, and provided high level overviews of plant status and production levels.

Origins

Early minicomputers were used in the control of industrial processes since the beginning of the 1960s. The IBM 1800, for example, was an early computer that had input/output hardware to gather process signals in a plant for conversion from field contact levels (for digital points) and analog signals to the digital domain.

The first industrial control computer system was built 1959 at the Texaco Port Arthur, Texas, refinery with an RW-300 of the Ramo-Wooldridge Company.

In 1975, both Honeywell and Japanese electrical engineering firm Yokogawa introduced their own independently produced DCS's - TDC 2000 and CENTUM systems, respectively. US-based Bristol also introduced their UCS 3000 universal controller in 1975. In 1978 Valmet introduced their own DCS system called Damatic (latest generation named Valmet DNA). In 1980, Bailey (now part of ABB) introduced the NETWORK 90 system, Fisher Controls (now part of Emerson Electric) introduced the PROVoX system, Fischer & Porter Company (now also part of ABB) introduced DCI-4000 (DCI stands for Distributed Control Instrumentation).

The DCS largely came about due to the increased availability of microcomputers and the proliferation of microprocessors in the world of process control. Computers had already been applied to process automation for some time in the form of both direct digital control (DDC) and setpoint control. In the early 1970s Taylor Instrument Company, (now part of ABB) developed the 1010 system, Foxboro the FOX1 system, Fisher Controls the DC^2 system and Bailey Controls the 1055 systems. All of these were DDC applications implemented within minicomputers (DEC PDP-11, Varian Data Machines, MODCOMP etc.) and connected to proprietary Input/Output hardware. Sophisticated (for the time) continuous as well as batch control was implemented in this way. A more conservative approach was setpoint control, where process computers supervised clusters of analog process controllers. A CRT-based workstation provided visibility into the process using text and crude character graphics. Availability of a fully functional graphical user interface was a way away.

Development

Central to the DCS model was the inclusion of control function blocks. Function blocks evolved from early, more primitive DDC concepts of "Table Driven" software. One of the first embodiments of object-oriented software, function blocks were self-contained "blocks" of code that emulated analog hardware control components and performed tasks that were essential to process control, such as execution of PID algorithms. Function blocks continue to endure as the predominant method of control for DCS suppliers, and are supported by key technologies such as Foundation Fieldbus today.

Midac Systems, of Sydney, Australia, developed an objected-oriented distributed direct digital control system in 1982. The central system ran 11 microprocessors sharing tasks and common memory and connected to a serial communication network of distributed controllers each running two Z80s. The system was installed at the University of Melbourne.

Digital communication between distributed controllers, workstations and other computing elements (peer to peer access) was one of the primary advantages of the DCS. Attention was duly focused on the networks, which provided the all-important lines of communication that, for process applications, had to incorporate specific functions such as determinism and redundancy. As a result, many suppliers embraced the IEEE 802.4 networking standard. This decision set the stage for the wave of migrations necessary when information technology moved into process automation and IEEE 802.3 rather than IEEE 802.4 prevailed as the control LAN.

The Network-Centric Era of the 1980s

In the 1980s, users began to look at DCSs as more than just basic process control. A very early example of a Direct Digital Control DCS was completed by the Australian business Midac in 1981–82 using R-Tec Australian designed hardware. The system installed at the University of Melbourne used a serial communications network, connecting campus buildings back to a control room "front end". Each remote unit ran two Z80 microprocessors, while the front end ran eleven Z80s in a parallel processing configuration with paged common memory to share tasks and that could run up to 20,000 concurrent control objects.

It was believed that if openness could be achieved and greater amounts of data could be shared throughout the enterprise that even greater things could be achieved. The first attempts to increase the openness of DCSs resulted in the adoption of the predominant operating system of the day: *UNIX*. UNIX and its companion networking technology TCP-IP were developed by the US Department of Defense for openness, which was precisely the issue the process industries were looking to resolve.

As a result, suppliers also began to adopt Ethernet-based networks with their own proprietary protocol layers. The full TCP/IP standard was not implemented, but the use of Ethernet made it possible to implement the first instances of object management and global data access technology. The 1980s also witnessed the first PLCs integrated into the DCS infrastructure. Plant-wide historians also emerged to capitalize on the extended reach of automation systems. The first DCS supplier to adopt UNIX and Ethernet networking technologies was Foxboro, who introduced the I/A Series system in 1987.

The Application-Centric Era of the 1990s

The drive toward openness in the 1980s gained momentum through the 1990s with the increased adoption of commercial off-the-shelf (COTS) components and IT standards. Probably the biggest transition undertaken during this time was the move from the UNIX operating system to the Windows environment. While the realm of the real time operating system (RTOS) for control applications remains dominated by real time commercial variants of UNIX or proprietary operating systems, everything above real-time control has made the transition to Windows.

The introduction of Microsoft at the desktop and server layers resulted in the development of technologies such as OLE for process control (OPC), which is now a de facto industry connectivity standard. Internet technology also began to make its mark in automation and the DCS world, with most DCS HMI supporting Internet connectivity. The 1990s were also known for the "Fieldbus Wars", where rival organizations competed to define what would become the IEC fieldbus standard for digital communication with field instrumentation instead of 4–20 milliamp analog communications. The first fieldbus installations occurred in the 1990s. Towards the end of the decade, the technology began to develop significant momentum, with the market consolidated around Ethernet I/P, Foundation Fieldbus and Profibus PA for process automation applications. Some suppliers built new systems from the ground up to maximize functionality with fieldbus, such as Rockwell PlantPAX System, Honeywell with Experion & Plantscape SCADA systems, ABB with System 800xA, Emerson Process Management with the Emerson Process Management DeltaV control system, Siemens with the SPPA-T3000 or Simatic PCS 7, Forbes Marshall with the Microcon+ control system and Azbil Corporation with the Harmonas-DEO system. Fieldbus technics have been used to integrate machine, drives, quality and condition monitoring applications to one DCS with Valmet DNA system.

The impact of COTS, however, was most pronounced at the hardware layer. For years, the primary business of DCS suppliers had been the supply of large amounts of hardware, particularly I/O and controllers. The initial proliferation of DCSs required the installation of prodigious amounts of this hardware, most of it manufactured from the bottom up by DCS suppliers. Standard computer components from manufacturers such as Intel and Motorola, however, made it cost prohibitive for DCS suppliers to continue making their own components, workstations, and networking hardware.

As the suppliers made the transition to COTS components, they also discovered that the hardware market was shrinking fast. COTS not only resulted in lower manufacturing costs for the supplier, but also steadily decreasing prices for the end users, who were also becoming increasingly vocal over what they perceived to be unduly high hardware costs. Some suppliers that were previously stronger in the PLC business, such as Rockwell Automation and Siemens, were able to leverage their expertise in manufacturing control hardware to enter the DCS marketplace with cost effective offerings, while the stability/ scalability/reliability and functionality of these emerging systems are still improving. The traditional DCS suppliers introduced new generation DCS System based on the latest Communication and IEC Standards, which resulting in a trend of combining the traditional concepts/functionalities for PLC and DCS into a one for all solution—named "Process Automation System". The gaps among the various systems remain at the areas such as: the database integrity, pre-engineering functionality, system maturity, communication transparency and reliability. While it is expected the cost ratio is relatively the same (the more powerful the systems are, the more expensive they will be), the reality of the automation business is often operating strategically case by case. The current next evolution step is called Collaborative Process Automation Systems.

To compound the issue, suppliers were also realizing that the hardware market was becoming saturated. The life cycle of hardware components such as I/O and wiring is also typically in the range of 15 to over 20 years, making for a challenging replacement market. Many of the older systems that were installed in the 1970s and 1980s are still in use today, and there is a considerable installed base of systems in the market that are approaching the end of their useful life. Developed industrial economies in North America, Europe, and Japan already had many thousands of DCSs installed, and with few if any new plants being built, the market for new hardware was shifting rapidly to smaller, albeit faster growing regions such as China, Latin America, and Eastern Europe.

Because of the shrinking hardware business, suppliers began to make the challenging transition from a hardware-based business model to one based on software and value-added services. It is a transition that is still being made today. The applications portfolio offered by suppliers expanded considerably in the '90s to include areas such as production management, model-based control, real-time optimization, plant asset management (PAM), Real-time performance management (RPM) tools, alarm management, and many others. To obtain the true value from these applications, however, often requires a considerable service content, which the suppliers also provide.

Modern Systems (2010 Onwards)

The latest developments in DCS include the following new technologies:

- Wireless systems and protocols
- Remote transmission, logging and data historian
- Mobile interfaces and controls
- Embedded web-servers

Increasingly, and ironically, DCS are becoming centralized at plant level, with the ability to log in to the remote equipment. This enables operator to control both at enterprise level (macro) and at the equipment level (micro) both within and outside the plant as physical location due to interconnectivity primarily due to wireless and remote access has shrunk.

As wireless protocols are developed and refined, DCS increasingly includes wireless communication. DCS controllers are now often equipped with embedded servers and provide on-the-go web access. Whether DCS will lead IIOT or borrow key elements from remains to be established.

Many vendors provide the option of a mobile HMI, ready for both Android and iOS. With these interfaces, the threat of security breaches and possible damage to plant and process are now very real.

REFERENCES

1. D'Andrea, Raffaello (9 September 2003). "Distributed Control Design for Spatially Interconnected Systems". IEEE Transactions on Automatic Control.
2. Massiaoni, Paolo (1 January 2009). "Distributed Control for Identical Dynamically Coupled Systems: A Decomposition Approach". IEEE Transactions on Automatic Control.
3. Stout, T. M.; Williams, T. J. (1995). "Pioneering Work in the Field of Computer Process Control". IEEE Annals of the History of Computing. 17 (1): 6–18. doi:10.1109/85.366507.
4. Valmet DNA
5. INFI 90
6. DCI-4000
7. Foundation Fieldbus
8. Foxboro I/A Series Distributed Control System
9. ABB System 800xA
10. Emerson Process Management
11. SPPA-T3000
12. "Archived copy". Archived from the original on 2007-03-29. Retrieved 2007-03-29. Simatic PCS 7
13. Forbes Marshall
14. Azbil Corporation

CHAPTER 10

Stochastic Control

Chapter Contents

Stochastic control refers to the general area in which some random variable distributions depend on the choice of certain controls, and one looks for an optimal strategy to choose those controls in order to maximize or minimize the expected value of the random variable.

The random variable to optimize is computed in terms of some stochastic process. It is usually the value of some given function evaluated at the end point of the stochastic process.

The expected value of this random variable, in terms of the starting point of the stochastic process, is a function which satisfies some fully nonlinear integro-differential equation.

INTRODUCTION

Stochastic control or stochastic optimal control is a subfield of control theory that deals with the existence of uncertainty either in observations or in the noise that drives the evolution of the system. The system designer assumes, in a Bayesian probability-driven fashion, that random noise with known probability distribution affects the evolution and observation of the state variables. Stochastic control aims to design the time path of the controlled variables that performs the desired control task with minimum cost, somehow defined, despite the presence of this noise. The context may be either discrete time or continuous time.

CERTAINTY EQUIVALENCE

An extremely well-studied formulation in stochastic control is that of linear quadratic Gaussian control. Here the model is linear, the objective function is the expected value of a quadratic form, and the disturbances are purely additive. A basic result for discrete-time centralized systems with only additive uncertainty is the certainty equivalence property: that the optimal control solution in this case is the same as would be obtained in the absence of the additive disturbances. This property is applicable to all centralized systems with linear equations of evolution, quadratic cost function, and noise entering the model only additively; the quadratic assumption allows for the optimal control laws, which follow the certainty-equivalence property, to be linear functions of the observations of the controllers.

Any deviation from the above assumptions—a nonlinear state equation, a non-quadratic objective function, noise in the multiplicative parameters of the model, or decentralization of control—causes the certainty equivalence property not to hold. For example, its failure to hold for decentralized control was demonstrated in Witsenhausen's counterexample.

DISCRETE TIME

In a discrete time context, the decision-maker observes the state variable, possibly with observational noise, in each time period. The objective may be to optimize the sum of expected values of a nonlinear (possibly quadratic) objective function over all the time periods from the present to the final period of concern, or to optimize the value of the objective function as of the final period only. At each time period new observations are made, and the control variables are to be adjusted optimally. Finding the optimal solution for the present time may involve iterating a matrix Riccati equation backwards in time from the last period to the present period.

In the discrete-time case with uncertainty about the parameter values in the transition matrix (giving the effect of current values of the state variables on their own evolution) and/or the control response matrix of the state equation, but still with a linear state equation and quadratic objective function, a Riccati equation can still be obtained for iterating backward to each period's solution even though certainty equivalence does not apply.[ch.13] The discrete-time case of a non-quadratic loss function but only additive disturbances can also be handled, albeit with more complications.

Example

A typical specification of the discrete-time stochastic linear quadratic control problem is to minimize

$$\mathrm{E}_1 \sum_{t=1}^{S} [y_t^T Q y_t + u_t^T R u_t]$$

where E_1 is the expected value operator conditional on y_0, superscript T indicates a matrix transpose, and S is the time horizon, subject to the state equation

$$y_t = A_t y_{t-1} + B_t u_t,$$

where y is an $n \times 1$ vector of observable state variables, u is a $k \times 1$ vector of control variables, A_t is the time t realization of the stochastic $n \times n$ state transition matrix, B_t is the time t realization of the stochastic $n \times k$ matrix of control multipliers, and Q ($n \times n$) and R ($k \times k$) are known symmetric positive definite cost matrices. We assume that the elements of A and B are jointly independently and identically distributed through time, so the expected value operations need not be time-conditional.

Induction backwards in time can be used to obtain the optimal control solution at each time,

$$u_t^* = -[\mathrm{E}(B^T X_t B + R)]^{-1} \mathrm{E}(B^T X_t A) y_{t-1},$$

with the symmetric positive definite cost-to-go matrix X evolving backwards in time from $X_S = Q$ according to

$$X_{t-1} = Q + \mathrm{E}[A^T X_t A] - \mathrm{E}[A^T X_t B][\mathrm{E}(B^T X_t B + R)]^{-1} \mathrm{E}(B^T X_t A),$$

which is known as the discrete-time dynamic Riccati equation of this problem. The only information needed regarding the unknown parameters in the A and B matrices is the expected value and variance of each element of each matrix and the covariances among elements of the same matrix and among elements across matrices.

The optimal control solution is unaffected if zero-mean, i.i.d. additive shocks also appear in the state equation, so long as they are uncorrelated with the parameters in the A and B matrices. But if they are so correlated, then the optimal control solution for each period contains an additional additive constant vector. If an additive constant vector appears in the state equation, then again the optimal control solution for each period contains an additional additive constant vector.

The steady-state characterization of X (if it exists), relevant for the infinite-horizon problem in which S goes to infinity, can be found by iterating the dynamic equation for X repeatedly until it converges; then X is characterized by removing the time subscripts from its dynamic equation.

CONTINUOUS TIME

If the model is in continuous time, the controller knows the state of the system at each instant of time. The objective is to maximize either an integral of, for example, a concave function of a state variable over a horizon from time zero (the present) to a terminal time T, or a concave function of a state variable at some future date T. As time evolves, new observations are continuously made and the control variables are continuously adjusted in optimal fashion.

In Finance

In a continuous time approach in a finance context, the state variable in the stochastic differential equation is usually wealth or net worth, and the controls are the shares placed at each time in the various assets. Given the asset allocation chosen at any time, the determinants of the change in wealth are usually the stochastic returns to assets and the interest rate on the risk-free asset. The field of stochastic control has developed greatly since the 1970s, particularly in its applications to finance. Robert Merton used stochastic control to study optimal portfolios of safe and risky assets. His work and that of Black–Scholes changed the nature of the finance literature. Influential mathematical textbook treatments were by Fleming and Rishel, and by Fleming and Soner. These techniques were applied by Stein to the financial crisis of 2007–08.

The maximization, say of the expected logarithm of net worth at a terminal date T, is subject to stochastic processes on the components of wealth. In this case, in continuous time Itô's equation is the main tool of analysis. In the case where the maximization is an integral of a concave function of utility over an horizon $(0,T)$, dynamic programming is used. There is no certainty equivalence as in the older literature, because the coefficients of the control variables—that is, the returns received by the chosen shares of assets—are stochastic.

STOCHASTIC PROCESS

A stochastic process (or random process) is a probability model used to describe phenomena that evolve over time or space. More specifically, in probability theory, a stochastic process is a time sequence representing the evolution of some system represented by a variable whose change is subject to a random variation. This is the probabilistic counterpart to a deterministic process (or deterministic system). Instead of describing a process which can only evolve in one way (as in the case, for example, of solutions of an ordinary differential equation), in a stochastic, or random process, there is some indeterminacy: even if the initial condition (or starting point) is known, there are several (often infinitely many) directions in which the process may evolve. In many stochastic processes, the movement to the next state or position depends on only the current state, and is independent from prior states or values the process has taken.

In the simple case of discrete time, as opposed to continuous time, a stochastic process is a sequence of random variables. (For example, see Markov chain, also known as discrete-time Markov chain.) The random variables corresponding to various times may be completely different, the only requirement being that these different random quantities all take values in the same space (the codomain of the function). One approach may be to model these random variables as random functions of one or several deterministic arguments (in most cases, the time parameter). Although the random values of a stochastic process at different times may be independent random variables, in most commonly considered situations they exhibit complicated statistical dependence.

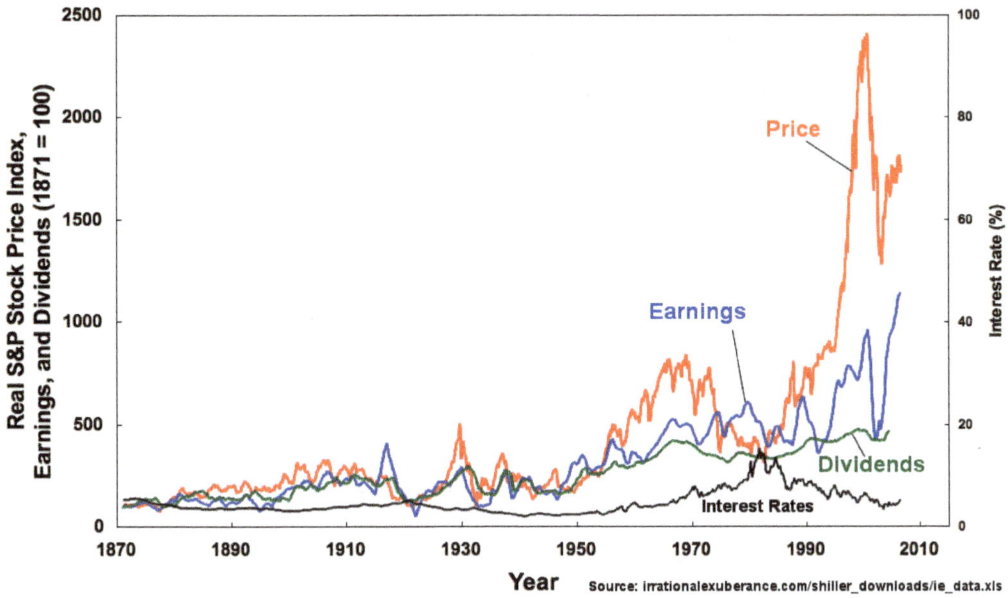

Stock market fluctuations have been modeled by stochastic processes.

Familiar examples of stochastic processes include stock market and exchange rate fluctuations; signals such as speech; audio and video; medical data such as a patient's EKG, EEG, blood pressure or temperature; and random movement such as Brownian motion or random walks.

A generalization, the random field, is defined by letting the variables be parametrized by members of a topological space instead of time. Examples of random fields include static images, random terrain (landscapes), wind waves and composition variations of a heterogeneous material.

Formal Definition and Basic Properties

Definition

Given a probability space $(\Omega, \mathcal{F}, P)$ and a measurable space (S, Σ), an S-valued **stochastic process** is a collection of S-valued random variables on , indexed by a totally ordered set T ("time"). That is, a stochastic process X is a collection

$$\{X_t : t \in T\}$$

where each X_t is an S-valued random variable on Ω. The space S is then called the **state space** of the process.

Finite-Dimensional Distributions

Let X be an S-valued stochastic process. For every finite sequence $T' = (t_1, \ldots, t_k) \in T^k$, the k-tuple $X_{T'} = (X_{t_1}, X_{t_2}, \ldots, X_{t_k})$ is a random variable taking values in S^k. The distribution $\mathbb{P}_{T'}(\cdot) = \mathbb{P}(X_{T'}^{-1}(\cdot))$ of this random variable is a probability measure on S^k. This is called a finite-dimensional distribution of X.

Under suitable topological restrictions, a suitably "consistent" collection of finite-dimensional distributions can be used to define a stochastic process.

History of Stochastic Processes

Stochastic processes were first studied rigorously in the late 19th century to aid in understanding financial markets and Brownian motion. The first person to describe the mathematics behind Brownian motion was Thorvald N. Thiele in a paper on the method of least squares published in 1880. This was followed independently by Louis Bachelier in 1900 in his PhD thesis "The theory of speculation", in which he presented a stochastic analysis of the stock and option markets. Albert Einstein (in one of his 1905 papers) and Marian Smoluchowski (1906) brought the solution of the problem to the attention of physicists, and presented it as a way to indirectly confirm the existence of atoms and molecules. Their equations describing Brownian motion were subsequently verified by the experimental work of Jean Baptiste Perrin in 1908.

An excerpt from Einstein's paper describes the fundamentals of a stochastic model:

"It must clearly be assumed that each individual particle executes a motion which is independent of the motions of all other particles; it will also be considered that the movements of one and the same particle in different time intervals are independent processes, as long as these time intervals are not chosen too small.

We introduce a time interval t into consideration, which is very small compared to the observable time intervals, but nevertheless so large that in two successive time intervals t, the motions executed by the particle can be thought of as events which are independent of each other".

Construction

In the ordinary axiomatization of probability theory by means of measure theory, the problem is to construct a sigma-algebra of measurable subsets of the space of all functions, and then put a finite measure on it. For this purpose one traditionally uses a method called Kolmogorov extension.

Kolmogorov Extension

The Kolmogorov extension proceeds along the following lines: assuming that a probability measure on the space of all functions $f : X \to Y$ exists, then it can be used to specify the joint probability distribution of finite-dimensional random variables $f(x_1), \ldots, f(x_n)$. Now, from this n-dimensional probability distribution we can deduce an $(n - 1)$-dimensional marginal probability distribution for $f(x_1), \ldots, f(x_{n-1})$. Note that the obvious compatibility condition, namely, that this marginal probability distribution be in the same class as the one derived from the full-blown stochastic process, is not a requirement. Such a condition only holds, for example, if the stochastic process is a Wiener process (in which case the marginals are all gaussian distributions of the exponential class) but not in general for all stochastic processes. When this condition is expressed in terms of probability densities, the result is called the Chapman–Kolmogorov equation.

The Kolmogorov extension theorem guarantees the existence of a stochastic process with a given family of finite-dimensional probability distributions satisfying the Chapman–Kolmogorov compatibility condition.

Separability, or what the Kolmogorov Extension does not Provide

Recall that in the Kolmogorov axiomatization, measurable sets are the sets which have a probability or, in other words, the sets corresponding to yes/no questions that have a probabilistic answer.

The Kolmogorov extension starts by declaring to be measurable all sets of functions where finitely many coordinates $[f(x_1), \ldots, f(x_n)]$ are restricted to lie in measurable subsets of Y^n. In other words, if a yes/no question about f can be answered by looking at the values of at most finitely many coordinates, then it has a probabilistic answer.

In measure theory, if we have a countably infinite collection of measurable sets, then the union and intersection of all of them is a measurable set. For our purposes, this means that yes/no questions that depend on countably many coordinates have a probabilistic answer.

The good news is that the Kolmogorov extension makes it possible to construct stochastic processes with fairly arbitrary finite-dimensional distributions. Also, every question that one could ask about a sequence has a probabilistic answer when asked of a random sequence. The bad news is that certain questions about functions on a continuous domain don't have a probabilistic answer. One might hope that the questions that depend on uncountably many values of a function be of little interest, but the really bad news is that virtually all concepts of calculus are of this sort. For example:

- boundedness
- continuity
- differentiability

all require knowledge of uncountably many values of the function.

One solution to this problem is to require that the stochastic process be separable. In other words, that there be some countable set of coordinates $\{f(x_i)\}$ whose values determine the whole random function f.

The Kolmogorov continuity theorem guarantees that processes that satisfy certain constraints on the moments of their increments have continuous modifications and are therefore separable.

Filtrations

Given a probability space $(\Omega, \mathcal{F}, P)$, a filtration is a weakly increasing collection of sigma-algebras on Ω, $\{\mathcal{F}_t, t \in T\}$, indexed by some totally ordered set T, and bounded above by $\mathcal{F}$, i.e. for $s,t \in T$ with $s < t$,

$$\mathcal{F}_s \subseteq \mathcal{F}_t \subseteq \mathcal{F}.$$

A stochastic process X on the same time set T is said to be adapted to the filtration if, for every $t \in T$, X_t is $\mathcal{F}_t$-measurable.

Classification

Stochastic processes can be classified according to the cardinality of its index set (usually interpreted as time) and state space.

Discrete Time and Discrete State Space

If both t and X_t belong to N, the set of natural numbers, then we have models which lead to Markov chains. For example:

(a) If X_t means the bit (0 or 1) in position t of a sequence of transmitted bits, then X_t can be modelled as a Markov chain with two states. This leads to the error-correcting Viterbi algorithm in data transmission.

(b) If X_t represents the combined genotype of a breeding couple in the tth generation in an inbreeding model, it can be shown that the proportion of heterozygous individuals in the population approaches zero as t goes to ∞.

Continuous Time and Continuous State Space

The paradigm of continuous stochastic process is that of the Wiener process. In its original form the problem was concerned with a particle floating on a liquid surface, receiving "kicks" from the molecules of the liquid. The particle is then viewed as being subject to a random force which, since the molecules are very small and very close together, is treated as being continuous and since the particle is constrained to the surface of the liquid by surface tension, is at each point in time a vector parallel to the surface. Thus, the random force is described by a two-component stochastic process; two real-valued random variables are associated to each point in the index set, time, (note that since the liquid is viewed as being homogeneous the force is independent of the spatial coordinates) with the domain of the two random variables being **R**, giving the *x* and *y* components of the force. A treatment of Brownian motion generally also includes the effect of viscosity, resulting in an equation of motion known as the Langevin equation.

Discrete Time and Continuous State Space

If the index set of the process is **N** (the natural numbers), and the range is **R** (the real numbers), there are some natural questions to ask about the sample sequences of a process $\{\mathbf{X}_i\}_{i \in \mathbf{N}}$, where a sample sequence is $\{\mathbf{X}_i(\omega)\}_{i \in \mathbf{N}}$.

- What is the probability that each sample sequence is bounded?
- What is the probability that each sample sequence is monotonic?
- What is the probability that each sample sequence has a limit as the index approaches ∞?
- What is the probability that the series obtained from a sample sequence from f(i) converges?
- What is the probability distribution of the sum?

Main applications of discrete time continuous state stochastic models include Markov chain Monte Carlo (MCMC) and the analysis of Time Series.

Continuous Time and Discrete State Space

Similarly, if the index space *I* is a finite or infinite interval, we can ask about the sample paths $\{\mathbf{X}_t(\omega)\}_{t \in I}$

- What is the probability that it is bounded/integrable...?
- What is the probability that it has a limit at ∞
- What is the probability distribution of the integral?

REFERENCES

1. Definition from Answers.com
2. Chow, Gregory P. (1976). Analysis and Control of Dynamic Economic Systems. New York: Wiley. ISBN 0-471-15616-7.
3. Turnovsky, Stephen (1976). "Optimal Stabilization Policies for Stochastic Linear Systems: The Case of Correlated Multiplicative and Additive disturbances". Review of Economic Studies. 43 (1): 191–94. doi:10.2307/2296614.
4. Mitchell, Douglas W. (1990). "Tractable Risk Sensitive Control Based on Approximate Expected Utility". Economic Modelling. 7 (2): 161–164. doi:10.1016/0264-9993(90)90018-Y.
5. Turnovsky, Stephen (1974). "The stability properties of optimal economic policies". American Economic Review. 64 (1): 136–148. JSTOR 1814888.
6. Merton, Robert (1990). Continuous Time Finance. Blackwell.
7. Fleming, W.; Rishel, R. (1975). Deterministic and Stochastic Optimal Control. ISBN 0-387-90155-8.
8. Fleming, W.; Soner, M. (2006). Controlled Markov Processes and Viscosity Solutions. Springer.
9. Stein, J. L. (2012). Stochastic Optimal Control and the US Financial Crisis. Springer-Science.

10. Dodge, Yadolah (2006). The Oxford Dictionary of Statistical Terms. Oxford, England: Oxford University Press. p. 335. ISBN 9780199206131.
11. Lindsey, J. K. (2004). Statistical Analysis of Stochastic Processes in Time. Cambridge, England: Cambridge University Press. p. 3. ISBN 9780521837415.
12. Lawler, G. (2006). Introduction to Stochastic processes (2 ed.). CRC Press. p. 1. A stochastic process is a random process evolving with time. More precisely, a stochastic process is a collection of random variables {\displaystyle X_{t}} X_{t} indexed by time.
13. Einstein, Albert (1926). "Investigations on the Theory of the Brownian Movement" (PDF).
14. Karlin, Samuel & Taylor, Howard M. (1998). An Introduction to Stochastic Modeling, Academic Press. ISBN 0-12-684887-4.
15. Durrett, Rick (2010). Probability: Theory and Examples (Fourth ed.). Cambridge: Cambridge University Press. ISBN 978-0-521-76539-8.
16. Allen, Linda J. S., An Introduction to Stochastic Processes with Applications to Biology, 2nd Edition, Chapman and Hall, 2010, ISBN 1-4398-1882-7
17. Gardiner, C. Handbook of Stochastic Methods: for Physics, Chemistry and the Natural Sciences, 3rd ed., Springer, 2004, ISBN 3540208828

Index